D1309951

Astronomy and Astrophysics Series Volume 10
Reference Works in Astronomy

Index of Galaxy Spectra

Astronomy and Astrophysics Series
General Editor: A.G. Pacholczyk

Journals

The Astronomy Quarterly
Edited by E.R. Craine

Index of Galaxy Spectra

Galen R. Gisler and Eileen D. Friel

Kitt Peak National Observatory

Pachart Publishing House

Tucson

Library of Congress Catalog Number: 79-87490
International Standard Book Number: 0-912918-19-5

Pachart Publishing House
1130 San Lucas Circle
Tucson, Arizona 85704

Table of Contents

Introduction

In spite of the proliferation of astronomical catalogues in recent years, there is still a notable lack of a bibliography of spectral information for galaxies. Existing galaxy catalogues, for example, the *Uppsala General Catalogue of Galaxies* (UGC, Nilson 1973), the *Second Reference Catalogue of Bright Galaxies* (2RC, de Vaucouleurs, de Vaucouleurs and Corwin 1976), the *Catalogue of Galaxies and Clusters of Galaxies* (CGCG, Zwicky et al. 1960-1968), and the *Morfologicheskii Katalog Galaktik* (MKG, Vorontsov-Velyaminov et al. (1962 - 1968) give radial velocities for many galaxies, but further information is often hard to find. The UGC and the CGCG give velocities without making any attempt to give references; the MKG gives velocities and often makes remarks as to the nature of the optical spectrum, but its coverage of the available spectral data is very spotty. The 2RC has the most complete collection of velocity data available up to now, and it gives references for all data. But the 2RC does not give further spectral information itself, and codes used for the bibliographical references do not pinpoint articles precisely, in most cases, so that the acquisition of further information is often a formidable task.

Because of the widespread interest in emission lines from galactic nuclei (with relevance both to the physical

1

nature of active galactic nuclei and to the interaction of an evolving galaxy with its environment) we have compiled this compendium of data on galaxy spectra and velocities from the astronomical literature, using the existing catalogues as guides. An earlier, unpublished version of this work has already been used in a statistical investigation of the effects of the intergalactic medium on the gas contents of cluster galaxies (Gisler 1978). The literature search for the present version is complete through August 1978 for *AJ*, *APJ*, *A&A*, *MNRAS*, and *PASP*.

We refer to this work as an index because we regard it fundamentally as a guide to the astronomical literature available on galaxy spectral data, and to this end we have tried to make our bibliographical references as precise and as complete as possible. While the Index may be used as a catalogue of galaxy redshifts, the production of such a catalogue has not been our primary aim. We have not, for example, attempted to pass judgment on the relative merits of different independent measurements of the radial velocity of the same galaxy, nor have we made an assessment of the velocity errors. We have listed all measurements which we have found, pertaining to a given galaxy. If the separate measurements have a standard deviation amongst themselves which exceeds 300 km s^{-1}, there may be a serious error (e.g., misidentification) in one of the measurements, and we indicate such a condition by the words "DISCREPANT VELOCITIES?" in the NOTES column. There are 50 galaxies in the Index to which this applies, but we have not attempted to ascertain which of the velocities given is more nearly correct.

The criteria for inclusion of a galaxy in the Index are:

(a) The galaxy must be listed in the UGC, from which are taken data on position, magnitude, Hubble type, and group or cluster membership.

(b) The galaxy must also have a published (through August 1978) radial velocity less than 15000 km s^{-1}. Independent measurements are included for each galaxy, and for each measurement the Index records the heliocentric radial velocity, an indication as to the nature of the spectrum (emission lines present - emission lines absent - no information), the spectral dispersion used, and the reference. Radial velocities obtained from observations of the 21-cm line of neutral hydrogen are also included to make the

2

information on velocities as complete as possible at present.

The Index so produced contains data for 2004 galaxies, of which 1603 have $m_p < 15.7$, $v_R < 15000$, spectral dispersions better than 500 Å/mm, and information as to the presence or absence of optical emission lines. Multiple velocities are available for 733 galaxies, and the total number of velocities listed in the Index is 3389. The largest number of velocities listed for a single galaxy is 15, for NGC 4151 (= UGC 7166).

In the appendices, data are listed for 432 clusters of galaxies, and 359 references are listed.

Explanation of the Index

The data for each galaxy are presented in the Index in two or more lines per galaxy, where the first line contains information about the galaxy gathered from the UGC (typographical errors which we noticed in the UGC have been corrected in the Index), and succeeding lines contain information about the spectrum of the galaxy, as published in the astronomical literature.

First line: (lefthand page)
UGC. The catalogue number given in the UGC. The suffix "A" refers to the addendum published at the end of the regular catalogue.

NGC. The NGC or IC (if prefaced by an "I") catalogue number of the galaxy. Suffixes such as "A" or "B" are not recognized in this Index, and therefore a single NGC number may refer to two or more entries.

RA. The 1950 right ascension as given in the UGC, to the nearest tenth of a minute, expressed in the form hh mm.m.

DEC. The 1950 declination as given in the UGC, to the nearest minute of arc, expressed in the format ± dd mm.

MAG. The magnitude as given in the UGC (usually from the CGCG).

TYPE. The galaxy's morphological type, from the UGC. The categories available are the Hubble types, with no distinction made between barred and unbarred types, plus "PEC" (peculiar), "MULT" (multiple), "DBLE" (double), "TRPL" (triple), "S..." (no attempt at classification).

CHAR. Distinguishing features noted in the UGC: "MARK" if the galaxy has been included in one of Markarian's (1967-1974) lists of galaxies with ultraviolet continua; "SEYF" if the galaxy's spectrum has Seyfert characteristics, according to one of its observers; "DISR" if the UGC describes the galaxy as "disrupted", "disturbed", "inter-acting", etc. For the first two categories, UGC information has been supplemented by reference to Markarian's more recent lists.

CLUSTER. If the galaxy lies within the contour boundary of a "near" distance class cluster in the CGCG, this column contains the serial number of that cluster given by the UGC. The special cluster number 500 has been assigned to galaxies which are considered likely Virgo cluster members. Otherwise this column is blank. These cluster numbers are identified with the better known cluster names in an appendix to the Index.

GROUP. If the galaxy is assigned to a group by the UGC, then this column contains the UGC number of the brightest galaxy in the group. Otherwise this column is blank. The number 99999 in this column identifies the galaxy as a probable member of the local group.

Underneath the cluster and group numbers will be printed the words "BR IN CLUS", "BR IN GRP", or "2 BR IN CLU", if the UGC remarks that the galaxy is the brightest in its cluster, brightest in its group, or second brightest in its cluster, respectively.

First line: (righthand page)
UGC. The UGC number from column 1 is repeated for convenience.

NOTES. Directly under the heading "VELOCITY" are printed the words "MULT VEL" if there exist more than one spectral/velocity measurement for the galaxy. Directly under the word "NOTES" are printed the words "MAG SHARED" if the magnitude given applies to the galaxy together with a near neighbor. "MAG NOT ZW" means that the magnitude given is not from the CGCG, but is either estimated by Nilson in the UGC or given by another source. Just to the left of this column is the catalogue number of the associated radio source, according to Dixon's "Master List of Radio Sources", but only if one of the spectral references contains explicit mention of an identification.

Continuation lines: (lefthand page)
TYPE. If the reference for the spectrum at the end of this line offers a morphological type differing from that given in the UGC, that type is given here, directly under the UGC type in the first line.

Continuation lines: (righthand page)
VELOCITY. This is the heliocentric velocity of recession of the galaxy in km s^{-1}, as given by the reference.
SPECTRUM. If the reference noted the presence of one or more emission lines in the spectrum, the symbol "E" appears here. If the reference did not remark that emission lines were present in this galaxy, but did so for other galaxies in the same paper or series of papers, or if the reference specifically indicated the absence of emission lines, then the symbol "A" appears. Finally, if there is no information whatsoever in the reference as to the presence or absence of optical emission lines, the symbol "N" appears.
DISPERSION. This is the inverse dispersion at which the spectrum was obtained, in eight possible ranges: 0-100 Å/mm, 101-200 Å/mm, 201-300 Å/mm, 301-400 Å/mm, 401-500 Å/mm, 501-700 Å/mm, 701-1000 Å/mm, or "VERY LOW", meaning > 1000 Å/mm. This information was usually available in the paper which contained the spectral data, although sometimes not immediately apparent. Phrases such as "low dispersion" without further explanation were roughly interpreted in the context of the earlier work of the author and his/her institute. If the author gave only a range of dispersions which overlapped two of our ranges, the spectrum was assigned to the poorer range. On the other hand, if the author presented evidence specifically based on a number of spectra at different dispersions, the best dispersion used is quoted. If there was no way of establishing, either directly or contextually, the spectral dispersion used, the words "NO INFO" appear. Finally, if the velocity given is obtained from radio data at 21 cm, the words "21-CM VELOCITY" appear in this column.
REFERENCE. This column contains a code of one or two letters, sometimes followed by a serial number, which

identifies the paper, or the short series of papers in which the data given in this line was published. The alphabetic part of the code is usually a mnemonic of some kind. These codes are precisely identified in a bibliographical appendix.

NOTES. In this column the continuation lines contain the words "VEL CONTAM" if the velocity quoted here refers to the galaxy together with a near neighbor, and not exclusively to the galaxy itself. The words "COMP VEL" appear if the UGC number actually refers to a double or multiple system of galaxies, and if the observer attempted to measure separate velocities for more than one component of the system. A one-symbol identifier (e.g., "A", "B", "N", "S") indicates the component to which the velocity applies.

Appendices to the Index

1. List of clusters:
    Position, Character, Population, and Radius (in
centimeters on the Palomar Sky Survey) are here taken from
the CGCG for all distance-class "near" clusters which contain any UGC galaxies. The cluster numbers are those given
by the UGC, and used in the body of the Index. Also given
are identifications of these clusters with clusters of common
names (such as Hercules, Perseus) and with clusters in the
Abell catalogue having distance classes less than 4. We
have assigned the cluster number 500 to the Virgo cluster,
which is not recognized by the CGCG or UGC. The character
class for this cluster, "medium compact", is taken from a
table in Zwicky's (1959) article in the *Handbuch der Physik*,
and the other parameters are provisional estimates. In
the body of the Index, a galaxy has been placed in the
Virgo cluster if any of the references for the spectrum, or
the notes of the UGC, attributed cluster membership to the
galaxy.

1. List of references:
    The reference codes used in the Index are here
identified with authors and journal references. The codes
"MW" and "L" refer to the separate lists of Mount Wilson
and Lick velocities published by Humason et al. 1956. The
book and journal names are abbreviated as follows:

8

ACTA	=	*Acta Astronomica*
ADA	=	*Annales D'Astrophysique*
AFZ	=	*Astrofizika*
AJ	=	*Astronomical Journal*
APJ	=	*Astrophysical Journal*
APLET	=	*Astrophysical Letters*
ATS	=	*Astronomicheskii Tsirkulyar*
AUJP	=	*Australian Journal of Physics*
AZH	=	*Astronomicheskii Zhurnal*
A&A	=	*Astronomy and Astrophysics*
CGCG	=	*Catalogue of Galaxies and Clusters of Galaxies,* F. Zwicky et al. 1960-1968, Cal. Inst. of Tech.
CKPNO	=	*Contributions of the Kitt Peak National Observatory*
CMRN	=	*Comptes Rendus*
COAA	=	*Contributi Dell'Osservatorio Astrofisica Dell' Università di Padova in Asiago*
CSCEG	=	*Catalogue of Selected Compact Galaxies and of Post Eruptive Galaxies,* F. Zwicky 1971.
IAUS	=	*IAU Symposium*
JOBS	=	*Journal des Observateurs*
MEM	=	*Memoirs of the Royal Astronomical Society*
MN	=	*Monthly Notices of the Royal Astronomical Society*
NAT	=	*Nature*
NG	=	*Study Week on Nuclei of Galaxies,* ed. by D. J. K. O'Connell, Elsevier 1971.
PASJ	=	*Publications of the Astronomical Society of Japan*
PASP	=	*Publications of the Astronomical Society of the Pacific*
SCI	=	*Science*
SSR	=	*Supernovae and Supernova Remnants,* ed. by C. B. Cosmovici, Dordrecht, 1974.
THES	=	Unpublished thesis
4BMSP	=	*Fourth Berkeley Symposium of Mathematical Statistics and Probability,* U. C. Press, 1961.

Where appropriate, these abbreviations are supplemented by the letters "S" (for supplement), "L" (for letters), or "P" (for *Monthly Notices* short contributions).

9

Gisler, ,G. R. 1978, *Mon. Not. Roy. Astr. Soc.* 183, 633.
Humason, M. L., Mayall, N. U., Sandage, A. R. 1956, *Astron. J.* 61, 97.
Nilson, P. N. 1973, *Uppsala General Catalogue of Galaxies* (UGC), Uppsala Obs. Ann., vol. 6.
Markarian, B. E. 1967-1969, *Astrofizika* 3, 55; 5, 443; 5, 481.
Markarian, B. E. and Lipovetsky, V. A. 1971-1974, *Astrofizika* 7, 511; 8, 155; 9, 487; 10, 307.
de Vaucouleurs, G., de Vaucouleurs, A., Corwin, 1976, *Second Reference Catalogue of Bright Galaxies* (2RC), Austin.
Vorontsov-Velyaminov, B. A., Krasnogorskaya, A. A., Arkhipova, V. P. 1962-1968, *Morfologicheskii Katalog Galaktik* (MKG), Moscow.
Zwicky, F. 1959, *Handbuch der Physik*, vol. LIII, ed. S. Flugge, Springer-Verlag, Berlin.
Zwicky, F. et al. 1960-1968, *Catalogue of Galaxies and Clusters of Galaxies* (CGCG), Cal. Inst. Tech., Pasadena.

Index of Galaxy Spectra

UGC	NGC	R.A. (1950) DEC	m_{pg}	Hubble Type	Char.	Cluster	Group
6		00 00.6 +21 42	14.4	PEC	MARK		
8	N7814	00 00.7 +15 52	12.0	SA-B			
16	N7816	00 01.2 +07 11	14.0	SB-C SA		BR[2] IN CLUS	
17		00 01.3 +14 56	17.0	SC IRR			
19	N7817	00 01.4 +20 28	12.7	SB-C			
26	N7819	00 01.8 +31 12	14.3	SB SC			
34	N7824	00 02.6 +06 38	14.5	SA-B		2	
36		00 02.7 +06 30	14.7	SA		2	
57	N 1	00 04.6 +27 26	13.4	SB			
58	N 3	00 04.7 +08 02	14.6	SO		2	
65	N7836	00 05.4 +32 48	13.8	...	MARK		
74	N 12	00 06.2 +04 20	14.5	SC			
80	N 16	00 06.4 +27 27	12.5	SO			
89	N 23	00 07.3 +25 39	12.5	SA SB			
106	N 36	00 08.8 +06 06	14.5	SB			
170	N 68	00 15.7 +29 48	14.5	SO		BR[4] IN GRP	170
173	N 71	00 15.8 +29 47	14.8	E		4	170
174	N 70	00 15.8 +29 49	14.5	SB		4	170
176	N 72	00 15.9 +29 46	15.0	SA		4	170
191		00 17.6 +10 36	15.6	SC-I			
192	I 10	00 17.6 +59 02	13.3	IRR SC			99999
193	N 78	00 17.8 +00 33	14.5	SO-A			
203	N 80	00 18.6 +22 05	13.7	SO		BR[5] IN GRP	203
206	N 83	00 18.8 +22 09	14.3	E		5	203
208	N 91	00 19.2 +22 08	14.5	S... SC	DISR	5	203
214	N 95	00 19.8 +10 13	13.4	SC SC-I			
226		00 21.2 +14 24	15.1	...	MARK		
230	N 99	00 21.4 +15 30	14.0	SC			

UGC	V_r (km s^{-1})	Spectrum	Disp. (Å/mm)	Ref.	Notes
6	6900	E	201-300	AD 5	
8	1047	A	401-500	MW	
16	5111	A	101-200	SN 8	
17	885	N	21-CM	FT	MAG NOT ZW
19	1837	A	301-400	V 8	
26	4858	E	101-200	FR11	
34	3119	N	101-200	TU	
36	3065	N	101-200	TU	
57	4628	E	301-400	KR	
58	3900	E	201-300	AD 9	
65	5100	E	201-300	AD 5	
74	3946	E	101-200	FR11	
80	3110	A	401-500	MW	
89	4568	E	401-500	MW	
106	6076	A	101-200	SN 8	
170 MULT VEL					
	5787	A	401-500	MW	
	5647	N	1-100	SA 5	
	5705	N	201-300	TG 3	
173 MULT VEL					
	6591	A	401-500	MW	
	6683	N	1-100	SA 5	
174 MULT VEL					
	7126	N	1-100	SA 5	
	7055	N	201-300	TG 3	
176 MULT VEL					
	6976	A	401-500	MW	
	6931	N	1-100	SA 5	
191	1144	N	21-CM	FT	
192 MULT VEL					MAG NOT ZW
	-343	E	201-300	MW	
	-346	N	21-CM	DD	
	-346	N	21-CM	S 1	
	-350	N	21-CM	RR	
	-341	N	21-CM	GU	
193	5250	E	301-400	DS 4	MAG SHARED
203	5586	A	401-500	MW	
206	6541	A	401-500	MW	
208	5168	N	101-200	FR11	
214	4886	A	301-400	V 5	
226	5400	E	201-300	AD 5	
230	5154	E	101-200	SN 8	

UGC	NGC	R.A. (1950) DEC	m_{pg}	Hubble Type	Char.	Cluster	Group
233		00 22.1 +14 33	14.6	PEC	MARK		
241	N 105	00 22.8 +12 37	14.1	SA SC			
262		00 24.6 +39 31	15.1	PEC	DISR		
279		00 25.7 +30 32	14.3	SB	MARK	6	
286	N 125	00 26.3 +02 33	14.2	SO		7	292
292	N 128	00 26.7 +02 35	13.2	SO	DISR	7 BR IN GRP	292
305		00 28.0 +13 05	15.3	SC	DISR		
312		00 28.8 +08 11	14.6	SB		BR IN GRP	312
326	N 147	00 30.4 +48 14	12.0	E			99999
356	N 160	00 33.4 +23 41	13.7	SA			
365	N 169	00 34.2 +23 43	13.3	SB	MARK		
367		00 34.4 +25 25	14.8	COMP		9	
369	N 173	00 34.6 +01 40	14.5	SB-C SC		7	
380	N 180	00 35.4 +08 22	14.3	SC			
382	N 182	00 35.7 +02 27	13.8	SA		7 BR IN GRP	382
386		00 35.8 +14 46	15.2	SC SO	MARK		
396	N 185	00 36.2 +48 03	11.0	E			99999
397	N 190	00 36.3 +06 46	14.4	SA-B		8	
402		00 36.7 +03 41	14.8	SO	MARK	7	
407	N 194	00 36.8 +02 45	13.9	COMP E		7	382
408	N 193	00 36.8 +03 03	14.3	COMP E		7	382
410	I1565	00 36.8 +06 27	14.9	COMP E		8	
420	N 200	00 37.1 +02 37	14.0	SC		7	382
426	N 205	00 37.6 +41 25	9.4	E SO			99999
438	N 214	00 38.8 +25 13	13.0	SC			
439		00 38.9 −01 59	14.4	SA			
448	I 43	00 39.7 +29 22	14.4	SC			

UGC	V_r (km s^{-1})	Spectrum	Disp. (A/mm)	Ref.	Notes
233					
	5400	E	201-300	AD 5	MAG SHARED
241					
	5260	E	101-200	SN 8	
262					
	10939	A	301-400	SA 2	
279					
	6081	N	201-300	HS	
286					
	5289	E	401-500	MW	
292	MULT VEL				MAG SHARED
	4250	E	401-500	MW	
	4255	E	NO INFO	BB38	
	4242	E	1-100	BP 1	
305					
	9904	A	101-200	SN 8	
312	MULT VEL				
	4341	N	101-200	TU	
	4200	E	101-200	DS 4	
326					
	-329	A	301-400	V 8	
356					
	5255	A	401-500	MW	
365	MULT VEL				MAG SHARED
	5487	E	301-400	KR	
	4500	E	201-300	AD 5	DISCREPANT VELOCITIES?
367					
	9630	A	NO INFO	UL17	
369	MULT VEL				
	4201	A	101-200	SN 8	
	4478	N	101-200	FR11	
380					
	5221	A	101-200	SN 8	
382	MULT VEL				
	5234	A	401-500	MW	
	5192	A	101-200	SN 8	
386	MULT VEL				
	5400	E	201-300	AD 5	
	5450	N	201-300	HS	
396	MULT VEL				
	-263	A	301-400	LB	
	-266	A	401-500	MW	
	-241	A	401-500	L	
397					MAG SHARED
	12107	E	201-300	CR 2	
402					
	12750	E	301-400	DS 4	
407					
	5105	A	401-500	MW	
408					4C03.01
	4350	A	201-300	WL	
410					
	11160	A	101-200	PT	
420					
	5110	E	101-200	SN 8	
426	MULT VEL				
	-239	A	201-300	MW	
	-233	A	401-500	L	
	-268	A	301-400	V 7	
438	MULT VEL				
	4482	A	401-500	L	
	4535	E	401-500	MW	
439					
	5100	E	201-300	BS	
448					
	4963	N	101-200	FR11	

15

UGC	NGC	R.A. (1950) DEC	m_{pg}	Hubble Type	Char.	Cluster	Group
452	N 221	00 39.9 +40 36	9.2	E			99999
454	N 224	00 40.0 +41 00	4.3	S B			99999
456	N 227	00 40.1 −01 48	13.7	E			
461	N 237	00 40.9 −00 24	13.6	SB-C SC			
468	I 49	00 41.4 +01 35	14.5	SC	DISR		
476	N 245	00 43.5 −01 59	12.9	S...	MARK		
486		00 44.6 +50 37	14.8	SB SC			
493	N 257	00 45.5 +08 03	13.7	SC			
499	N 262	00 46.1 +31 42	15.0	SO	SEYF		
528	N 278	00 49.2 +47 17	10.5	S... SB			
532	N 279	00 49.7 −02 28	14.0	SO	MARK		
540		00 50.2 +28 45	14.1	PEC		20	
579		00 53.7 −01 30	15.0	E		14 2BR IN CLU	
583		00 53.9 −01 30	14.9	E		14 BR IN CLUS	
587		00 54.4 −01 28	15.3	SC		14	
591		00 54.6 +23 37	15.2	...	MARK		
594	N 317	00 54.8 +43 31	13.8	S B	DISR		
595		00 55.1 −01 38	14.8	COMP E E		14	
597	N 315	00 55.1 +30 05	12.5	E		20	
600		00 55.5 +48 23	14.5	SB SC			
601	N 326	00 55.7 +26 36	14.9	DBLE E		16	
602		00 55.7 +36 28	14.5	SC			
634		00 58.8 +07 21	15.7	SC-I			

UGC	V_r (km s^{-1})	Spectrum	Disp. (A/mm)	Ref.	Notes
452	MULT VEL				
	-193	A	401-500	L	
	-214	A	1-100	MW	
	-259	A	301-400	V 7	
	-300	N	NO INFO	SB	
454	MULT VEL				
	-266	A	1-100	MW	
	-290	A	401-500	L	
	-300	N	NO INFO	SB	
	-314	E	1-100	DH 1	
	-300	E	1-100	FR 7	
	-300	E	101-200	FR 6	
	-300	N	21-CM	BK	
	-289	N	21-CM	DR	
	-303	N	21-CM	DG	
	-296	N	21-CM	DD	
	-310	N	21-CM	RW	
456					
	5315	A	401-500	MW	
461					
	4109	A	101-200	SN 8	
468					
	4571	E	101-200	FR11	
476	MULT VEL				
	4350	E	301-400	DS 4	
	4084	E	101-200	SN 8	
486					
	5206	E	101-200	FR11	
493					
	5272	A	101-200	SN 8	
499	MULT VEL				
	4200	E	201-300	AD 5	
	4453	N	21-CM	BI 1	
528	MULT VEL				
	622	E	201-300	MW	
	656	E	401-500	L	
	650	N	NO INFO	SB	
	655	N	21-CM	BA 1	
532					
	4050	E	301-400	DS 4	
540					
	5100	E	201-300	AD 9	
579					
	13200	A	101-200	PT	
583					
	11509	A	101-200	SN 8	
587					
	15018	E	101-200	SN 8	
591					
	5100	E	201-300	AD 5	
594					MAG SHARED
	5324	E	301-400	KR	
595	MULT VEL				3C029
	13350	A	101-200	T	
	13384	A	301-400	SN 8	
597					
	5010	A	NO INFO	UL17	
600					
	6903	N	101-200	FR11	
601	MULT VEL				4C26.03
	14490	A	101-200	SA 4	
	14160	A	NO INFO	UL17	
602					
634	6127	E	101-200	FR11	
	2211	N	21-CM	FT	

UGC	NGC	R.A. (1950) DEC	m_{pg}	Hubble Type	Char.	Cluster	Group
645	N 354	01 00.6 +22 05	14.2	S B	MARK		
646		01 00.7 +31 58	15.0	S B		20	
668	I1613	01 02.4 +01 53	10.7	IRR			99999
671	I1618	01 03.2 +32 09	15.6	SO		20	
680	N 374	01 04.3 +32 32	14.3	SO-A		20	
682	N 380	01 04.5 +32 13	13.9	E		20	689
683	N 379	01 04.5 +32 15	14.0	SO		20	689
686	N 384	01 04.7 +32 01	14.3	SO		20	689
687	N 385	01 04.7 +32 03	14.3	E		20	689
688	N 382	01 04.7 +32 08	14.2	E		20	689
689	N 383	01 04.7 +32 09	13.6	SO		20 BR IN GRP	689
697		01 05.3 +33 11	14.7	S B		20	
700	N 392	01 05.6 +32 52	13.9	E-SO		20 BR IN GRP	700
701		01 05.7 +01 56	15.5	DBLE E		19	
710		01 06.0 +33 12	15.6	SB-C		20	
711		01 06.1 +01 23	14.8	SC		19	
712	N 399	01 06.2 +32 22	14.5	S A		20	
715	N 403	01 06.5 +32 29	13.3	SO-A S...		20	
718	N 404	01 06.7 +35 27	11.3	E-SO		18	
724		01 07.2 +32 05	14.0	S...		20	
730	N 407	01 07.8 +32 51	14.3	SO-A S...		20	
731		01 07.8 +49 21	17.0	IRR		22	
735	N 410	01 08.2 +32 53	12.6	E		20	
743		01 08.5 +31 37	14.8	S A		20	
749		01 09.0 +01 04	14.2	SC-I		19	
750	I1639	01 09.2 -00 55	14.2	...	MARK	19	
752	N 420	01 09.3 +31 52	13.4	SO		20	
763	N 428	01 10.4 +00 43	11.9	... SC		19	
773		01 11.2 +02 07	16.0	...			

UGC	V_r (km s^{-1})	Spectrum	Disp. (A/mm)	Ref.	Notes
645	MULT VEL				
	4800	E	201-300	AD 5	
	4845	N	101-200	HS	
646	5418	A	201-300	DM	
668	-238	E	201-300	MW	
671	4706	A	201-300	DM	
680	5067	A	201-300	DM	
682	4341	A	401-500	MW	
683	5374	A	401-500	MW	
686	MULT VEL				
	4401	A	401-500	MW	
	4500	E	201-300	AD 9	
687	4845	A	401-500	MW	
688	5156	A	401-500	MW	3C031
689	MULT VEL				3C031
	4888	A	401-500	MW	
	4984	E	301-400	KR	
697	4719	E	201-300	DM	
700	4672	A	201-300	DM	
701	13110	A	101-200	PT	
710	12624	E	201-300	DM	
711	2010	N	NO INFO	BS	
712	5167	A	201-300	DM	
715	4977	A	201-300	DM	
718	MULT VEL				
	-55	E	201-300	MW	
	14	E	401-500	LB	
	-25	N	NO INFO	SB	
	-26	E	NO INFO	BB38	
724	5414	A	201-300	DM	
730	5610	A	201-300	DM	
731	646	N	21-CM	FT	MAG NOT ZW
735	5238	A	201-300	DM	
743	5229	A	201-300	DM	
749	6900	E	201-300	AD 9	
750	2550	E	301-400	DS 4	
752	5199	A	201-300	DM	
763	MULT VEL				
	1078	E	401-500	L	
	1130	N	21-CM	GU	
	1175	N	21-CM	R 3	
773	14100	E	1-100	WL	MAG NOT ZW PKS0111+02

UGC	NGC	R.A. (1950) DEC	m_{pg}	Hubble Type	Char.	Cluster	Group
776	N 431	01 11.3 +33 27	14.0	SO		20	
783		01 11.5 +42 17	14.6	SC			
792	I1652	01 12.2 +31 42	14.3	SO-A		20	
793		01 12.3 −00 45	14.5	DBLE SC			
797		01 12.4 +00 10	15.5	E		19	
804	N 449	01 12.8 +32 48	14.0	SO-A	SEYF	20	
806	N 450	01 13.0 −01 07	13.0	SC			
818	I 89	01 13.5 +04 02	13.8	SC-I SO	MARK		
833		01 15.5 +11 07	14.3	SC			
842		01 16.4 −01 16	15.4	...		25	
848	N 467	01 16.6 +03 02	13.3	SO			858
855		01 17.0 +07 55	14.9	SB SC			
858	N 470	01 17.2 +03 08	12.4	SB-C		BR IN GRP	858
859	N 473	01 17.2 +16 17	13.2	SO SA			
864	N 474	01 17.5 +03 09	12.9	SO	DISR		858
891		01 18.6 +12 09	15.7	IRR			
894	I1681	01 18.8 −00 10	14.8	S...	DISR	25	
907	N 488	01 19.2 +05 00	11.6	SB			
912	I1682	01 19.4 +33 00	14.3	...			
913		01 19.4 +34 25	14.4	PEC			
915	N 497	01 19.8 −01 08	14.1	SB-C SC		25	
919	N 494	01 20.1 +32 55	13.8	SA-B		20	
920	N 495	01 20.1 +33 13	14.0	SO-A		20	
921		01 20.2 −01 39	14.6	S...		25	
926	N 499	01 20.4 +33 13	13.0	SO		20	
928		01 20.6 −00 54	15.2	SO		25	
929		01 20.6 −00 39	14.9	SC		25	
935	N 504	01 20.7 +32 57	14.0	SO		20	
938	N 507	01 20.8 +33 00	13.0	E		20	
939	N 508	01 20.8 +33 02	14.5	E		20	

UGC	V_r (km s^{-1})	Spectrum	Disp. (A/mm)	Ref.	Notes
776	5786	A	201-300	DM	
783	5914	E	101-200	FR11	
792	5317	A	201-300	DM	
793	10113	E	101-200	SN 8	
797	13439	A	101-200	FD	
804	MULT VEL				
	4800	E	1-100	WD 1	
	4800	E	101-200	WK 1	
806	MULT VEL				MAG SHARED
	1761	N	21-CM	S 3	
	1884	E	101-200	SN 8	
818	6450	E	301-400	DS 4	
833	5032	A	101-200	SN 8	
842	13478	A	NO INFO	CR 6	
848	5568	A	301-400	V 5	
855	9439	E	101-200	SN 8	
858	MULT VEL				
	2557	E	301-400	V 5	
	2332	E	101-200	SN 8	
859	2222	A	101-200	SN 8	
864	2306	A	401-500	MW	
891	643	N	21-CM	FT	
894	3889	E	NO INFO	CR 6	
907	2180	A	401-500	MW	
912	18900	E	201-300	AD 9	
913	5400	E	201-300	AD 9	
915	MULT VEL				
	8030	N	301-400	ZH 3	
	8110	A	101-200	SN 8	
919	5513	E	201-300	TI 4	
920	4114	A	401-500	MW	
921	5651	N	301-400	ZH 3	
926	4375	A	401-500	MW	
928	8021	A	NO INFO	CR 6	
929	7617	E	NO INFO	CR 6	
935	4288	A	201-300	TI 4	
938	MULT VEL				
	4929	A	401-500	MW	
	4935	A	201-300	TI 4	
939	5674	A	201-300	TI 4	

UGC	NGC	R.A. (1950) DEC	m pg	Hubble Type	Char.	Cluster	Group
940		01 20.8 +34 19	14.9	SC		20	
947	N 514	01 21.4 +12 39	12.8	SC			
953	N 513	01 21.7 +33 33	13.4	...		20	
959		01 21.9 +31 55	14.2	SA	SEYF	20	
962	N 521	01 22.0 +01 28	12.9	SB SC		23 BR IN CLUS	
965	N 530	01 22.1 −01 50	14.0	SA		25	
966	N 520	01 22.1 +03 33	12.4	PEC IRR			
968	N 524	01 22.1 +09 17	11.5	SO		24	
973	I1696	01 22.3 −01 52	14.7	E		25	
974		01 22.3 −01 45	15.0	SO-A		25	
976	I1697	01 22.5 +00 10	14.9	...		23	
979	N 523	01 22.5 +33 46	13.5	PEC		20	
983	I1698	01 22.6 +14 35	14.9	SO			986
984		01 22.7 −01 46	14.8	SO		25	
986	I1700	01 22.7 +14 36	14.3	E	DISR		986 BR IN GRP
991	N 538	01 22.9 −01 48	14.7	SA		25	
992	N 533	01 22.9 +01 29	13.1	E		23	
996		01 23.0 −01 45	14.8	SO-A		25	
997	N 535	01 23.0 −01 39	14.9	SO		25	
1003		01 23.2 −01 42	15.0	SO		25	
1004	N 541	01 23.2 −01 37	14.0	E		25	1009
1007	N 545	01 23.4 −01 35	13.7	SO		25	1009
1009	N 547	01 23.5 −01 36	13.4	E		25 BR IN CLUS	1009
1010	N 548	01 23.5 −01 29	15.1	E-SO		25	
1013	N 536	01 23.6 +34 27	13.2	SB		20	

22

UGC	V_r (km s^{-1})	Spectrum	Disp. (A/mm)	Ref.	Notes
940	MULT VEL				
	6852	E	301-400	SA 2	
	6870	E	NO INFO	Z 3	
947	MULT VEL				
	2487	A	401-500	MW	
	2602	A	401-500	L	
953					
	6000	E	201-300	AD 9	
959					
	10800	E	201-300	AD 9	
962					
	4969	A	101-200	SN 8	
965					
	5016	N	101-200	ZH 3	
966	MULT VEL				
	2084	A	401-500	L	
	2751	A	301-400	LB	COMP VEL S
	2326	E	301-400	LB	COMP VEL N
	1860	E	1-100	KH 1	COMP VEL S
	2070	E	1-100	KH 1	COMP VEL N
	2280	N	21-CM	DL 2	
	2260	N	21-CM	PS	
968					
	2470	A	401-500	MW	
973					
	5768	N	101-200	ZH 3	
974					
	4788	N	301-400	ZH 3	
976					
	8707	E	NO INFO	CR 6	
979	MULT VEL				
	4701	E	201-300	CR 2	COMP VEL W
	4762	E	201-300	CR 2	COMP VEL E
983					
	5365	N	101-200	TU	
984					
	5118	N	101-200	ZH 3	
986					
	5747	N	101-200	TU	
991	MULT VEL				
	5398	N	101-200	ZH 3	
	5107	A	NO INFO	CR 6	
992	MULT VEL				
	5003	A	301-400	V 5	
	5476	A	101-200	SN 8	
				DISCREPANT VELOCITIES?	
996	MULT VEL				
	6591	N	101-200	ZH 3	
	5577	A	NO INFO	CR 6	
				DISCREPANT VELOCITIES?	
997					
	4939	N	101-200	ZH 3	
1003					
	5321	N	301-400	ZH 3	
1004					
	5392	N	101-200	ZH 3	
1007	MULT VEL				3C040
	5499	A	301-400	V 5	
	5316	N	101-200	ZH 3	
1009	MULT VEL				3C040
	5361	A	301-400	V 5	
	5472	N	101-200	ZH 3	
1010					
	5332	N	101-200	ZH 3	
1013					
	5200	E	201-300	BC	

UGC	NGC	R.A. (1950) DEC	m$_{pg}$	Hubble Type	Char.	Cluster	Group
1016	I1703	01 23.8 −01 54	14.9	S0−A		25	
1027	I1706	01 24.5 +14 31	14.2	S... SC			
1030		01 24.7 −01 31	14.9	E		25	
1032		01 24.8 +18 55	13.8	PEC	MARK	26	
1036	N 560	01 24.9 −02 10	14.0	S0		25	
1040		01 25.0 −01 21	14.8	S0−A		25	
1043		01 25.1 −01 23	15.1	E		25	
1044	N 564	01 25.2 −02 08	13.8	E		25	
1047	I 119	01 25.3 −02 17	15.0	S0−A		25	
1049	N 562	01 25.4 +48 07	14.5	SC		22	
1052	N 565	01 25.6 −01 33	14.5	SA		25	
1055		01 26.0 −01 59	15.1	SA		25	
1061	N 570	01 26.4 −01 12	14.2	S0−A		25	
1062		01 26.4 −00 49	14.0	S...		25	
1071	I 126	01 27.2 −02 14	15.7	...		25	
1072		01 27.2 −01 30	14.7	S0		25	
1080	N 577	01 28.1 −02 15	14.2	SA		25	
1092	N 585	01 29.1 −01 11	14.2	SA		25	
1095		01 29.3 +31 51	15.0	S...	DISR	20	
1106	I 138	01 30.4 −00 57	14.9	SC		25	
1117	N 598	01 31.0 +30 24	6.5	SC			99999
1123		01 31.6 −01 17	14.4	SA−B		25 BR IN GRP	1123
1133		01 32.5 +04 06	16.5	IRR			
1143	N 622	01 33.4 +00 25	14.1	SB	MARK		

UGC	V_r (km s^{-1})	Spectrum	Disp. (A/mm)	Ref.	Notes
1016					
	5688	N	101-200	ZH 3	
1027					
	6319	E	101-200	SN 8	
1030					
	4794	N	301-400	ZH 3	
1032					
	5100	E	201-300	AD 5	
1036	MULT VEL				
	5503	A	401-500	MW	
	5405	N	101-200	ZH 3	
	5454	N	NO INFO	Z 2	
1040					
	4661	N	301-400	ZH 3	
1043					
	5174	N	301-400	ZH 3	
1044	MULT VEL				
	5851	A	401-500	MW	
	5556	N	101-200	ZH 3	
	5704	N	NO INFO	Z 2	
1047	MULT VEL				
	6203	N	101-200	ZH 3	
	5712	A	NO INFO	CR 6	DISCREPANT VELOCITIES?
1049					
	10268	E	101-200	FR11	
1052					
	4464	N	101-200	ZH 3	
1055					
	6271	N	301-400	ZH 3	
1061					
	5502	N	101-200	ZH 3	
1062					
	5333	A	NO INFO	CR 6	
1071					
	5712	A	NO INFO	CR 6	
1072	MULT VEL				
	900	E	201-300	AD 9	
	5137	A	NO INFO	CR 6	DISCREPANT VELOCITIES?
1080					
	6057	E	NO INFO	CR 6	
1092					
	5211	A	NO INFO	CR 6	
1095	MULT VEL				
	12577	E	201-300	CR 2	COMP VEL W
	12542	E	201-300	CR 2	COMP VEL E
1106					
	4588	E	NO INFO	CR 6	
1117	MULT VEL				
	-178	E	1-100	LS	
	-189	A	101-200	MW	
	-195	N	101-200	BJ	
	-172	E	1-100	CN	
	-180	N	21-CM	WW	
	-170	N	21-CM	ME	
	-181	N	21-CM	DJ	
	-184	N	21-CM	DD	
	-180	N	21-CM	HU 1	
1123					
	4934	A	NO INFO	CR 6	
1133					MAG NOT ZW
	1966	N	21-CM	FT	
1143					
	5400	E	301-400	DS 4	

25

UGC	NGC	R.A. (1950) DEC	m_{pg}	Hubble Type	Char.	Cluster	Group
1149	N 628	01 34.0 +15 32	10.5	SC			
1153	N 631	01 34.2 +05 35	15.0	E			
1157	N 632	01 34.7 +05 37	13.5	SO			
1167		01 35.8 +07 16	14.8	SC			
1176		01 37.4 +15 39	17.0	IRR			
1192	N 658	01 39.5 +12 20	13.6	SB SC		29	
1201	N 660	01 40.3 +13 23	12.8	SA	DISR	29	
1209		01 41.1 +11 55	15.0	SC-I	MARK	29	
1210	N 664	01 41.2 +03 59	13.9	SB SC		BR IN GRP	1210
1214		01 41.4 +02 06	14.0	SO	MARK		
1238	N 668	01 43.4 +36 12	13.5	SB			
1248	N 669	01 44.3 +35 18	12.9	SA-B S...		31	
1249	I1727	01 44.6 +27 05	12.2	... SC			1256
1250	N 670	01 44.6 +27 38	13.1	SO SA		32	
1256	N 672	01 45.0 +27 11	11.4	SC		BR IN GRP	1256
1259	N 673	01 45.7 +11 17	13.3	SC		29	
1260		01 45.8 +12 21	14.0	SA SC	MARK	29	
1266	I 162	01 46.1 +10 15	14.2	... S...		29 BR IN GRP	1266
1267		01 46.2 +10 16	14.8	SO		29	1266
1272		01 46.3 +34 50	14.3	SO SO		31 BR IN GRP	1272
1277		01 46.5 +35 12	14.5	SO-A S...		31	
1282		01 46.8 +12 15	14.2	SO-A	SEYF	29	
1283	N 679	01 46.8 +35 32	13.1	E-SO E		31	
1298	N 687	01 47.6 +36 07	13.3	SO E		31	

UGC	V_r (km s^{-1})	Spectrum	Disp. (A/mm)	Ref.	Notes
1149	MULT VEL				
	561	E	401-500	MW	
	652	E	1-100	DH 3	
	654	N	21-CM	DD	
	655	N	21-CM	RR	
	655	N	21-CM	GU	
1153	4010	N	101-200	TU	
1157	3310	N	101-200	TU	
1167	4199	E	101-200	SN 8	
1176	634	N	21-CM	FT	MAG NOT ZW
1192	2955	E	101-200	SN 8	
1201	1003	A	301-400	V 8	
1209	5250	E	101-200	DS 4	
1210	5382	E	101-200	SN 8	
1214	5100	E	301-400	DS 4	
1238	5151	A	201-300	DM	
1248	4756	A	201-300	DM	
1249	362	E	401-500	MW	
1250	MULT VEL				
	3189	A	301-400	V 8	
	3758	E	101-200	SN 8	
					DISCREPANT VELOCITIES?
1256	MULT VEL				
	340	E	401-500	L	
	420	N	21-CM	RR	
	562	N	21-CM	KS	
1259	5211	E	101-200	SN 8	
1260	MULT VEL				
	6150	E	301-400	DS 4	
	5265	E	101-200	SN 8	
					DISCREPANT VELOCITIES?
1266	5019	N	101-200	TU	
1267	4811	N	101-200	TU	
1272	MULT VEL				
	4864	A	201-300	DM	
	17400	E	201-300	AD 9	
					DISCREPANT VELOCITIES?
1277	4238	A	201-300	DM	
1282	MULT VEL				
	12600	E	101-200	DS 3	
	11850	E	301-400	DS 4	
					DISCREPANT VELOCITIES?
1283	MULT VEL				
	4853	A	201-300	DM	
	5026	A	101-200	FD	
1298	MULT VEL				
	5094	A	201-300	DM	
	5147	A	101-200	FD	

UGC	NGC	R.A. (1950) DEC	m_{pg}	Hubble Type	Char.	Cluster	Group
1302	N 688	01 47.8 +35 02	13.3	SB		31	
1307	I1732	01 47.9 +35 40	15.1	S...		31	
1308		01 47.9 +36 01	14.5	E		31	
1310	N 694	01 48.2 +21 45	13.9	PEC	MARK		
1325		01 48.9 +08 00	14.2	E			
1326		01 49.0 +08 03	15.0	E			
1334	N 706	01 49.2 +06 03	13.2	SC			1270
1336	N 700	01 49.2 +35 50	15.6	SO		31	
1338		01 49.4 +35 33	15.2	SB		31	
1343	N 704	01 49.7 +35 52	14.1	DBLE E		31	1348
1344		01 49.7 +36 15	14.0	SA		31	
1345	N 705	01 49.8 +35 55	14.5	SO-A SO		31	1348
1346	N 703	01 49.8 +35 56	14.5	E-SO S...		31	1348
1347		01 49.8 +36 22	13.9	SC S...		31	
1348	N 708	01 49.9 +35 55	14.8	E S...		31 BR IN GRP	1348
1350		01 50.0 +36 15	14.5	SB		31	
1351	I1743	01 50.2 +12 28	14.0	SA SC			
1352	N 712	01 50.2 +36 34	13.9	SO S...		31	
1353		01 50.4 +36 43	14.4	E-SO E		31	
1356	N 718	01 50.6 +03 57	12.5	SA			
1358	N 714	01 50.6 +35 58	13.9	SO-A SO		31	
1363	N 717	01 51.0 +35 59	14.7	SO-A S...		31	
1371	I1746	01 51.7 +04 33	15.1	SO			
1385		01 52.1 +36 41	14.2	SA S...	MARK	31 BR IN GRP	1385

UGC	V_r (km s^{-1})	Spectrum	Disp. (A/mm)	Ref.	Notes
1302					
	4096	E	201-300	DM	
1307					
	4889	A	101-200	FD	
1308	MULT VEL				
	5034	A	201-300	DM	
	5312	A	101-200	FD	
1310	MULT VEL				
	3000	E	201-300	AD	5
	2800	N	101-200	HS	
1325					
	5146	N	101-200	TU	
1326					
	4992	N	101-200	TU	
1334					
	4881	E	101-200	SN	8
1336					
	4364	A	201-300	DM	
1338					
	4099	A	101-200	FD	
1343	MULT VEL				
	4618	A	201-300	DM	
	4941	A	201-300	MS	
1344	MULT VEL				
	4428	A	201-300	DM	
	4398	A	101-200	FD	
1345	MULT VEL				
	4645	A	201-300	DM	
	4526	A	101-200	FD	
1346	MULT VEL				
	5175	A	201-300	DM	
	4800	A	NO INFO	UL17	
	5592	A	101-200	FD	
	4663	A	201-300	HS	DISCREPANT VELOCITIES?
1347					
	4996	A	201-300	DM	
1348	MULT VEL				
	4830	E	101-200	PT	
	5047	A	201-300	DM	
	4827	A	101-200	FD	
	4939	A	201-300	MS	
1350					
	5244	A	101-200	FD	
1351					
	4531	E	101-200	SN	8
1352	MULT VEL				
	5303	A	201-300	DM	
	5286	A	101-200	FD	
1353					
	5055	A	201-300	DM	
1356					
	1802	A	401-500	L	
1358	MULT VEL				
	4534	A	201-300	DM	
	4470	A	101-200	FD	
	4280	A	201-300	MS	
1363					
	4968	A	201-300	DM	
1371					
	780	N	NO INFO	BB44	
1385	MULT VEL				
	5400	E	101-200	WK	1
	5386	A	301-400	V	8
	5458	E	201-300	DM	
	5400	E	NO INFO	DS	2

UGC	NGC	R.A. (1950) DEC	m pg	Hubble Type	Char.	Cluster	Group
1388	I 171	01 52.3 +35 02	13.8	S...		31	
1395		01 52.7 +06 21	14.5	SB SC			
1402	I 173	01 53.4 +01 02	14.9	SB SC			
1406	N 732	01 53.5 +36 34	14.9	S0 S...		31	
1413	N 741	01 53.8 +05 23	13.2	E		BR IN GRP	1413
1414	N 736	01 53.8 +32 48	13.6	E		32	
1415		01 53.8 +36 08	14.5	S0-A		31	
1430	N 750	01 54.6 +32 58	12.9	E		32	
1431	N 751	01 54.6 +32 58	12.9	E		32	
1437	N 753	01 54.8 +35 40	12.6	SC		31	
1440	N 759	01 54.9 +36 05	13.7	E		31	
1449		01 55.6 +02 50	14.0	IRR	MARK		
1463	M 770	01 56.4 +18 43	14.2	S...			
1466	N 772	01 56.5 +18 46	11.3	SB			
1467	N 769	01 56.7 +30 40	13.4	S...		32	
1475	I 179	01 57.2 +37 47	13.4	E		31	
1476	N 777	01 57.3 +31 10	12.7	E		32	
1493		01 57.9 +37 58	14.0	SA-B S...		31	
1498		01 58.3 +08 04	14.4	S... SC			
1510		01 58.9 +26 18	14.4	...			
1520	N 789	01 59.5 +31 50	14.0	PEC		32	
1546		02 00.6 +18 24	14.8	SC		33	
1547		02 00.6 +21 48	15.0	IRR			
1550	N 801	02 00.7 +38 01	13.5	SC S...		31	
1555	I 195	02 01.0 +14 28	14.3	S0	DISR	35	
1556	I 196	02 01.1 +14 30	14.2	SB	DISR	35	
1592	I 198	02 03.4 +09 03	14.8	S... SC			
1631	N 821	02 05.7 +10 45	12.6	E			
1636	N 825	02 05.9 +06 05	14.5	SA			

UGC	V_r (km s^{-1})	Spectrum	Disp. (Å/mm)	Ref.	Notes
1388					
1395	5362	A	201-300	DM	
1402	5160 MULT VEL	E	101-200	SN 8	
	13877	E	101-200	SN 8	
	13916	E	101-200	FR11	
1406					
1413	5813	E	201-300	DM	
1414	5559	A	401-500	MW	
1415	4366	A	401-500	MW	
1430	4796	A	201-300	DM	
1431	5130	A	201-300	MW	MAG SHARED
1437	5126 MULT VEL	A	401-500	MW	MAG SHARED
	4766	A	401-500	L	
	4902	E	201-300	DM	
1440	4714	A	201-300	DM	
1449	6000 MULT VEL	E	301-400	DS 4	MAG SHARED
	2477	N	101-200	TU	
	2454	N	NO INFO	A 5	
1463					
1466	2431 MULT VEL	A	401-500	MW	
	2458	N	101-200	TU	
	2437	N	101-200	A 5	
	2435	N	21-CM	DU	
	2430	N	21-CM	R 3	
1467	4500	E	201-300	AD 9	
1475	4062	A	201-300	DM	
1476	5000 MULT VEL	A	301-400	V 8	
	4989	E	101-200	SN 8	
1493	4249	E	201-300	DM	
1498	4726	E	101-200	SN 8	
1510	5100	E	201-300	AD 9	
1520	5100	E	201-300	AD 9	
1546	2377	E	101-200	FR11	
1547	2635	N	21-CM	FT	
1550	5716	A	201-300	DM	
1555	8600	N	101-200	TU	
1556	8563	N	101-200	TU	
1592	9384	E	101-200	SN 8	
1631	1778	A	401-500	MW	
1636	3121	N	101-200	TU	

UGC	NGC	R.A. (1950) DEC	m_{pg}	Hubble Type	Char.	Cluster	Group
1651		02 06.7 +35 34	14.9	E	DISR	37	
1655	N 828	02 07.1 +38 57	13.0	PEC			
1664	N 840	02 07.6 +07 36	14.7	SB SC			
1669		02 08.0 +05 38	14.5	S...	MARK		
1670		02 08.0 +06 31	16.0	SC-I			
1672	N 834	02 08.0 +37 26	13.2	SB			37
1678	I 211	02 08.5 +03 38	14.5	SC			
1680	N 851	02 08.6 +03 33	14.7	SO	MARK		
1716		02 11.1 +03 53	14.3	COMP	MARK		
1727	N 863	02 12.0 -01 00	14.0	SA	SEYF		
1736	N 864	02 12.8 +05 46	12.0	SC			
1757		02 14.3 +38 11	13.6	S...	SEYF		
1759	N 871	02 14.4 +14 19	13.6	... SC		35	1768
1768	N 877	02 15.2 +14 19	12.5	SC		35 BR IN GRP	1768
1794		02 17.2 -00 29	14.6	SB	MARK		
1810		02 18.4 +39 09	13.7	SB	DISR		
1813		02 18.5 +39 08	15.3	SB	DISR		
1823	N 890	02 19.0 +33 03	12.5	SO		38	
1831	N 891	02 19.4 +42 07	10.8	SB		47	2154
1840		02 20.0 +41 09	14.1	PEC		47	
1841		02 20.0 +42 46	15.0	E		47	
1865		02 22.0 +35 49	16.5	SC-I			
1868	N 906	02 22.1 +41 52	14.4	SA SC		47	
1875	N 910	02 22.3 +41 36	14.5	E		47	
1880	I1797	02 22.7 +20 10	15.3	SB			
1888	N 918	02 23.1 +18 16	14.3	SC			
1901	N 926	02 23.6 -00 33	13.9	SB SC			

UGC	V_r (km s^{-1})	Spectrum	Disp. (A/mm)	Ref.	Notes
1651	MULT VEL				4C35.03
	11070	A	301-400	T	
	11190	A	101-200	SA 4	
1655	5430	E	NO INFO	UL17	
1664	7113	A	101-200	SN 8	
1669	4650	E	301-400	DS 4	
1670	1611	N	21-CM	FT	MAG NOT ZW
1672	4800	E	201-300	AD 9	
1678	MULT VEL				
	3075	N	101-200	TU	
	3288	E	101-200	SN 8	
	3236	E	101-200	FR11	
1680	MULT VEL				
	3450	E	301-400	DS 4	
	3199	N	101-200	TU	
	3045	E	301-400	KR	
1716	MULT VEL				
	4320	E	301-400	SA 2	
	3300	E	301-400	DS 4	
	3530	E	201-300	UL15	
	3364	E	201-300	BR 2	
1727					DISCREPANT VELOCITIES?
1736	8250	E	301-400	DS 4	
1757	1583	E	401-500	L	
1759	6000	E	201-300	AD 9	
	MULT VEL				
	3705	E	301-400	LB	
	3757	E	401-500	L	
1768	4016	A	301-400	L	
1794	7500	E	301-400	DS 4	
1810	7623	E	301-400	KR	
1813	7778	E	301-400	KR	
1823	4043	A	401-500	MW	
1831	72	A	401-500	MW	
1840	MULT VEL				
	5239	E	301-400	SA 2	
	5306	E	301-400	A 3	
1841	MULT VEL				MAG SHARED 3C066
	6450	A	NO INFO	MM	
	6706	N	NO INFO	A 3	
1865	MULT VEL				MAG NOT ZW
	575	N	21-CM	FT	
	592	N	21-CM	BA 4	
1868	4586	N	101-200	FR11	
1875	5130	A	101-200	PT	
1880	3900	E	201-300	AD 9	
1888	1502	E	101-200	FR11	
1901	6485	A	101-200	SN 8	

UGC	NGC	R.A. (1950) DEC	m pg	Hubble Type	Char.	Cluster	Group
1908	N 927	02 23.9 +11 56	14.5	SC			
1913	N 925	02 24.3 +33 21	10.5	SC			2154
1929	N 936	02 25.1 −01 23	11.3	S0 SA			
1936	I1801	02 25.4 +19 21	14.8	SB			
1937	N 935	02 25.4 +19 22	13.9	SC			
1954	N 941	02 26.0 −01 22	13.4	SC SC−I			
1983	N 949	02 27.8 +36 55	12.0	S...		37	
1986	N 955	02 28.0 −01 20	13.0	SC−I SA−B			
1995		02 28.8 +01 07	14.6	SB SC		39	
2010		02 29.6 −01 35	14.3	SB SC	DISR		
2014		02 29.8 +38 28	17.0	IRR			
2016	I 235	02 30.0 +20 25	14.5	PEC	MARK		
2023		02 30.3 +33 17	14.9	IRR		40	
2034		02 30.6 +40 19	15.0	IRR		47	
2042	N 976	02 31.2 +20 45	12.9	SB SB−C		38	
2045	N 972	02 31.3 +29 05	12.1	SA−B SB		38	
2053		02 31.5 +29 32	15.7	IRR		38	
2069		02 32.5 +37 25	13.2	SC		37	
2079		02 33.3 +23 41	14.8	SC		38	
2080	I 239	02 33.3 +38 45	12.1	SC			2154
2103	N 992	02 34.6 +20 53	13.5	S...	MARK	38	
2105		02 34.6 +34 14	13.9	SA		40	
2122		02 35.5 +29 32	14.7	SC		38	
2132	N1019	02 35.8 +01 42	14.6	SB SC		39	

UGC	V_r (km s^{-1})	Spectrum	Disp. (A/mm)	Ref.	Notes
1908					
	8252	E	101-200	FR11	
1913	MULT VEL				
	587	E	401-500	L	
	420	E	401-500	MW	
	580	E	301-400	BB32	
	557	N	21-CM	DD	
	555	N	21-CM	RR	
	565	N	21-CM	GU	
	693	N	21-CM	KS	
	546	E	101-200	SN 8	
1929	MULT VEL				
	1343	E	401-500	MW	
	1300	N	NO INFO	SB	
1936	MULT VEL				
	4323	E	301-400	KR	
	4013	E	201-300	BM	
1937					
	4191	E	201-300	BM	
1954					
	1580	E	101-200	SN 8	
1983	MULT VEL				
	622	E	NO INFO	DS 5	
	563	E	101-200	SN 8	
1986					
	1504	A	101-200	SN 8	
1995					
	7360	E	101-200	SN 8	
2010					
	11239	A	101-200	SN 8	
2014					MAG NOT ZW
	567	N	21-CM	FT	
2016					
	9000	E	201-300	AD 5	
2023					
	615	N	21-CM	FT	
2034					
	581	N	21-CM	FT	
2042					
	4332	A	101-200	SN 8	
2045	MULT VEL				
	1538	E	401-500	MW	
	1593	A	301-400	LB	
	1545	E	301-400	BB40	
	1544	E	NO INFO	UL 1	
2053	MULT VEL				
	1035	N	21-CM	FT	
	1034	N	21-CM	BA 4	
2069					
	3665	N	101-200	SA 4	
2079					
	5616	E	101-200	FR11	
2080					
	903	N	21-CM	S 3	
2103	MULT VEL				
	4500	E	201-300	AD 9	
	3642	E	201-300	BR 1	
	3900	E	NO INFO	AD 5	
	4135	N	21-CM	CX	
					DISCREPANT VELOCITIES?
2105	MULT VEL				
	4800	A	201-300	MW	
	5100	E	201-300	AD 9	
2122					
	5068	E	101-200	FR11	
2132					
	7221	E	101-200	SN 8	

UGC	NGC	R.A. (1950) DEC	m_{pg}	Hubble Type	Char.	Cluster	Group
2137	N1003	02 36.1 +40 40	12.1	SC		47	2154
2140		02 36.3 +18 10	14.6	IRR	DISR		
2142	N1024	02 36.5 +10 38	13.8	SB		BR IN GRP	2142
2149	N1029	02 36.9 +10 35	14.9	SO-A			2142
2154	N1023	02 37.2 +38 51	10.5	SO		BR IN GRP	2154
2160	N1036	02 37.6 +19 05	13.5	PEC	MARK		
2162		02 37.8 +01 01	18.0	IRR		39	
2173	N1055	02 39.1 +00 13	12.5	SB			2188
2188	N1068	02 40.1 -00 13	9.7	SB	SEYF	BR IN GRP	2188
2193	N1058	02 40.3 +37 08	11.8	SC		47	2154
2204	N1067	02 40.8 +32 18	14.6	SC		40	
2210	N1073	02 41.1 +01 10	12.5	SC			2188
2232		02 43.0 +36 42	16.0	∴ E		42	
2241	N1085	02 43.8 +03 24	13.6	SB			
2245	N1087	02 43.9 -00 42	11.4	SC		43	2188
2247	N1090	02 44.0 -00 27	12.8	SB SC		43	2188
2262	N1094	02 44.9 -00 30	13.5	SA-B SC		43	2188
2275		02 45.4 +03 41	15.7	SC-I			
2296		02 46.4 +18 07	13.1	COMP			
2302		02 46.6 +01 56	15.3	SC-I			
2345		02 49.3 -01 22	16.0	SC		43	

UGC	V_r (km s^{-1})	Spectrum	Disp. (A/mm)	Ref.	Notes
2137					
	585	E	401-500	MW	
2140					MAG SHARED
	4037	E	401-500	L	
2142					
	3572	N	101-200	TU	
2149					
	3513	N	101-200	TU	
2154	MULT VEL				
	557	A	201-300	MW	
	734	A	401-500	LB	
	600	N	21-CM	GU	
	590	N	21-CM	PS	
	670	N	21-CM	BI 2	
2160	MULT VEL				
	900	E	201-300	AD 5	
	748	E	301-400	BG 6	
	761	E	101-200	HS	
	787	N	21-CM	BG 6	
2162					MAG NOT ZW
	1186	N	21-CM	FT	
2173	MULT VEL				
	987	A	301-400	V 8	
	1050	N	21-CM	GU	
2188	MULT VEL				3C071
	1020	E	201-300	MW	
	1121	E	401-500	L	
	1203	E	301-400	BB 3	
	1107	E	1-100	W	
	1080	E	1-100	MC	
	1120	N	NO INFO	SB	
	1073	E	NO INFO	DA 1	
	1114	N	101-200	DA 2	
	1133	N	21-CM	DL 2	
	1145	N	21-CM	AL 2	
2193	MULT VEL				
	480	E	301-400	LS	
	521	E	301-400	V 5	
	439	N	301-400	Z 1	
	517	N	21-CM	DL 2	
2204					
	4535	E	101-200	FR11	
2210	MULT VEL				
	1239	E	101-200	LS	
	1209	E	101-200	FR 8	
	1218	N	21-CM	FR 8	
2232	MULT VEL				MAG NOT ZW
	14460	A	101-200	PT	
	14700	N	NO INFO	SN 5	
2241					
	6950	A	101-200	SN 8	
2245	MULT VEL				
	1536	E	301-400	V 7	
	1824	A	401-500	MW	
2247					
	2699	A	101-200	SN 8	
2262					
	6264	E	101-200	SN 8	
2275					
	1030	N	21-CM	FT	
2296					
	10010	A	201-300	UL15	
2302					
	1108	N	21-CM	FT	
2345					MAG NOT ZW
	1508	N	21-CM	FT	

37

UGC	NGC	R.A. (1950) DEC	m_{pg}	Hubble Type	Char.	Cluster	Group
2455	N1156	02 56.8 +25 03	12.0	IRR			
2460	I 277	02 57.2 +02 34	13.8	SB-C	MARK		
2487	N1167	02 58.6 +35 01	14.0	SO		47	
2489		02 58.7 +35 39	16.0	SC MULT			
2503	N1169	03 00.1 +46 12	13.2	SB		47	
2515	N1175	03 01.3 +42 09	13.8	S... SO-A		47	
2548	N1207	03 05.0 +38 11	13.7	SB		47	
2555	N1218	03 05.8 +03 55	14.0	SO-A			
2567	I 292	03 07.0 +40 35	14.3	SC-I		47	
2578	N1224	03 08.0 +41 11	15.5	E-SO		47	
2586	N1233	03 09.3 +39 08	13.9	SB		47	
2595	I 302	03 10.2 +04 31	14.0	SB-C SC			
2613	N1250	03 12.0 +41 10	14.2	SO		47	
2619		03 13.0 +41 00	16.5	SO		47	
2621		03 13.2 +41 21	14.7	SA		47	
2624	I 310	03 13.4 +41 08	14.3	SO		47	
2626		03 13.7 +41 10	15.7	SA		47	
2627		03 13.9 +31 23	14.9	SC	DISR		
2634	N1260	03 14.2 +41 13	14.2	SO-A	DISR	47	
2643	N1264	03 14.7 +41 20	16.0	SA-B		47	
2644	I 312	03 14.8 +41 34	14.9	E		47	
2651	N1265	03 15.1 +41 40	14.7	E		47	
2657	N1267	03 15.5 +41 17	15.4	E		47	
2658	N1268	03 15.5 +41 18	14.5	SB		47	
2660	N1270	03 15.7 +41 17	14.4	E		47	
2662	N1272	03 16.1 +41 18	14.5	E SO		47	
2665		03 16.2 +41 27	15.5	SC		47	
2669	N1275	03 16.5 +41 20	13.0	PEC IRR E	SEYF	47 BR IN CLUS	
2670	N1278	03 16.7 +41 22	14.4	E	DISR	47	

UGC	V_r (km s^{-1})	Spectrum	Disp. (A/mm)	Ref.	Notes
2455	MULT VEL				
	405	E	401-500	MW	
	380	N	21-CM	R 3	
	335	N	21-CM	BI 1	
2460	2700	E	301-400	DS 4	
2487	MULT VEL				4C34.09
	5100	E	101-200	WL	
	4800	E	101-200	SA 4	
2489	MULT VEL				MAG NOT ZW 4C35.06
	14070	A	101-200	PT	
	13980	A	101-200	SA 4	
2503	2342	E	101-200	SN 8	
2515	5428	A	101-200	SN 8	
2548	4769	E	301-400	KR	
2555	8640	E	301-400	ST	3C078
2567	3018	E	201-300	CR 1	
2578	5051	A	201-300	CR 1	
2586	4890	E	201-300	WL	4C39.11
2595	5905	N	21-CM	FR11	
2613	6197	A	201-300	CR 1	
2619	6292	A	201-300	CR 1	MAG NOT ZW
2621	4747	A	201-300	CR 1	
2624	MULT VEL				MW 0313+41
	5209	A	201-300	CR 1	
	5756	N	NO INFO	A 3	
2626					DISCREPANT VELOCITIES?
2627	6358	A	201-300	CR 1	
2634	4201	N	101-200	FR11	
2643	5521	A	201-300	CR 1	
2644	3246	A	201-300	CR 1	MAG NOT ZW
2651	4803	A	201-300	CR 1	
2657	7536	A	201-300	CR 1	3C083.1
2658	5111	A	201-300	CR 1	
2660	3124	A	201-300	CR 1	
2662	4905	A	401-500	MW	
2665	4172	A	301-400	LB	
2669	7861	E	201-300	CR 1	
	MULT VEL				3C084
	5160	E	201-300	MW	
	5265	E	101-200	BB41	
	5300	E	201-300	AS	
2670	6115	A	401-500	MW	

UGC	NGC	R.A. (1950) DEC	m pg	Hubble Type	Char.	Cluster	Group
2673		03 16.8 +41 04	15.6	SO		47	
2675	N1282	03 17.0 +41 11	14.3	E		47	
2676	N1283	03 17.0 +41 13	15.6	COMP		47	
2682	I 313	03 17.6 +41 42	15.1	E		47	
2683	N1298	03 17.7 −02 17	14.2	E		49	
2694	N1294	03 18.5 +41 11	15.1	SO		47	
2748		03 25.3 +02 23	15.5	E			2744
2783		03 31.0 +39 12	14.2	E		47	
2792	N1343	03 32.4 +72 24	14.1	PEC			
2821	N1409	03 38.7 −01 27	14.7	DBLE	SEYF		
2836		03 40.6 +39 09	13.8	E−SO		47 BR IN GRP	2836
2847	I 342	03 42.0 +67 57	10.5	SC			
2855		03 43.3 +70 00	14.6	SC			
2947	N1507	04 01.9 −02 19	13.1	SC−I			
2953	I 356	04 02.6 +69 41	13.3	SB−C			
2989		04 10.7 +29 03	16.0	S...			
3042		04 23.2 +70 15	15.6	SB−C			
3056	N1569	04 26.1 +64 45	11.8	SC IRR			
3060	N1560	04 27.1 +71 47	12.1	SC−I			
3063	N1587	04 28.1 +00 33	13.3	E		59	
3064	N1588	04 28.2 +00 33	14.1	COMP	DISR	59	
3071	N1590	04 28.5 +07 31	14.6	PEC			
3080		04 29.3 +01 05	14.9	SC		59	
3087		04 30.5 +05 15	14.2	SO COMP	SEYF		
3128		04 37.8 +04 06	15.3	SO SA			

UGC	V_r (km s^{-1})	Spectrum	Disp. (A/mm)	Ref.	Notes
2673					
2675	4219	A	201-300	CR 1	
2676	2203	A	201-300	CR 1	
2682	6727	A	201-300	CR 1	
2683	4429	A	201-300	CR 1	
	MULT VEL				
	300	E	201-300	AD 9	
	6630	N	NO INFO	BS	
					DISCREPANT VELOCITIES?
2694					
2748	6550	A	201-300	CR 1	
					3C088
	9060	E	301-400	SA 3	
2783	MULT VEL				
					4C39.12
	6270	A	201-300	WL	
	6060	A	NO INFO	UL17	
2792					
	300	A	NO INFO	F	
2821	MULT VEL				
	7402	E	301-400	SA 2	COMP VEL A
	7606	E	301-400	KR	COMP VEL B
2836					
	4950	E	101-200	PT	
2847	MULT VEL				
	8	E	101-200	FR 8	
	-10	E	201-300	MW	
	34	E	401-500	L	
	25	N	21-CM	RS	
2855					
					4CP69.05A
	1200	E	201-300	UL16	
2947					
	898	E	301-400	V 5	
2953	MULT VEL				
	743	N	21-CM	LS	
	870	N	21-CM	PS	
2989					
					MAG NOT ZW
	5239	A	301-400	SA 2	
3042					
	2947	N	101-200	FR11	
3056	MULT VEL				
	-34	E	201-300	MW	
	-58	E	401-500	L	
	-90	N	21-CM	RR	
	-95	N	21-CM	R 3	
3060	MULT VEL				
	-303	A	301-400	V 8	
	-43	N	21-CM	DL 2	
3063					
	3890	A	401-500	MW	
3064					
	3328	A	301-400	SA 2	
3071	MULT VEL				
	3768	E	301-400	SA 2	
	3927	N	21-CM	CX	
3080					
	3482	E	101-200	FR11	
3087	MULT VEL				
					4C05.20
	10020	E	301-400	BU	
	9966	E	301-400	SA 2	
	9750	E	201-300	WL	
	9900	E	1-100	OK	
	9900	E	NO INFO	A 2	
3128					
	4600	A	401-500	MW	

41

UGC	NGC	R.A. (1950) DEC	m_{pg}	Hubble Type	Char.	Cluster	Group
3133	N1638	04 39.0 -01 55	13.6	E-S0			
3144		04 41.4 +74 50	15.7	IRR			
3154	N1654	04 43.3 -02 11	14.4	S... E		62A	
3174		04 46.0 +00 10	17.0	IRR			
3179		04 47.2 +03 15	14.8	COMP			
3190	I 391	04 49.5 +78 07	12.8	S... SB			
3192		04 50.2 +01 10	14.6	E-S0		63 BR IN GRP	3192
3199		04 51.7 +01 35	15.5	SB		63	3198
3203	I 396	04 52.8 +68 14	13.2	S...			
3224		04 56.6 +05 33	14.4	SB SC			
3234		05 00.5 +16 20	16.5	IRR			
3271		05 13.6 -00 12	14.6	...	SEYF		
3274		05 14.0 +06 23	14.5	MULT E		68	
3302		05 21.0 +76 37	14.9	SC			
3317		05 27.5 +73 42	17.0	IRR		64	
3334	N1961	05 36.6 +69 21	12.2	SB			
3371		05 49.8 +75 19	17.0	IRR			
3373		05 51.0 +78 30	15.0	SC		76	
3393		06 00.4 +07 50	14.5	COMP			
3422		06 09.3 +71 09	14.5	SB SC			
3426		06 09.8 +71 03	13.8	S0	SEYF		
3429	N2146	06 10.5 +78 22	11.1	SA PEC		76	
3460		06 21.5 +74 18	15.0	SB	MARK		
3471		06 25.0 +74 27	14.9	SC			
3546	N2273	06 45.6 +60 54	12.5	SA	MARK		
3547	I 450	06 45.6 +74 28	14.8	S0-A	SEYF	85	
3573		06 48.7 +27 33	15.2	SB		88	
3581		06 50.0 +80 04	14.4	SC			

UGC	V_r (km s^{-1})	Spectrum	Disp. (A/mm)	Ref.	Notes
3133					
	3276	A	101-200	SN 8	
3144					
	1635	N	21-CM	FT	
3154					
	4577	E	201-300	BC	
3174					MAG NOT ZW
	669	N	21-CM	FT	
3179	MULT VEL				
	8377	E	301-400	SA 2	
	8184	E	301-400	KR	
	8429	E	201-300	BR 2	
3190					
	1607	E	401-500	L	
3192					
	17700	E	201-300	AD 9	
3199					
	2100	E	201-300	AD 9	
3203					
	762	E	301-400	KR	
3224					
	4734	N	101-200	FR11	
3234					MAG NOT ZW
	1390	N	21-CM	FT	
3271	MULT VEL				
	9900	E	201-300	AD 9	
	7050	E	101-200	OP	DISCREPANT VELOCITIES?
3274					MAG NOT ZW
	8100	A	101-200	PT	COMP VEL S
3302					
	4173	E	101-200	FR11	
3317					MAG NOT ZW
	1241	N	21-CM	FT	
3334					
	3870	E	401-500	L	
3371					MAG NOT ZW
	818	N	21-CM	FT	
3373					
	4821	N	101-200	FR11	
3393	MULT VEL				
	5286	A	301-400	SA 2	
	5560	A	NO INFO	Z 3	
3422					
	3960	E	101-200	FR11	
3426	MULT VEL				
	4050	E	101-200	WK 1	
	3930	E	1-100	WD 1	
	4050	E	NO INFO	DS 2	
3429	MULT VEL				
	711	E	301-400	BB 5	
	785	E	401-500	MW	
	784	E	401-500	L	
	856	N	21-CM	BA 2	
3460					
	4800	E	1-100	WK 1	
3471					
	5568	E	101-200	FR11	
3546					
	1950	E	101-200	DS 3	
3547	MULT VEL				
	5400	E	101-200	WK 1	
	5760	E	1-100	WD 1	
	5610	E	1-100	UL11	
3573					
	12270	E	NO INFO	UL17	
3581					
	4986	E	101-200	FR11	

UGC	NGC	R.A. (1950) DEC	m pg	Hubble Type	Char.	Cluster	Group
3647		07 00.6 +56 36	16.0	IRR		96	
3653	N2268	07 01.0 +84 28	12.1	SC		102	
3677	N2314	07 03.7 +75 25	13.1	E-S0 E			
3690		07 05.0 +53 33	17.0	IRR		96	
3693	N2339	07 05.4 +18 51	12.3	SB-C SB			
3695	N2329	07 05.4 +48 42	13.7	E-S0		89	
3697		07 05.6 +71 55	13.1	S...	DISR		
3708	N2341	07 06.3 +20 40	13.7	PEC			
3709	N2342	07 06.4 +20 43	12.6	S...	DISR		
3730		07 08.2 +73 33	13.7	PEC			
3734	N2344	07 08.7 +47 15	13.2	SB SC		89	
3740	N2276	07 10.0 +85 51	12.3	SC	DISR	102	
3759	N2347	07 11.3 +64 49	13.2	SB			
3798	N2300	07 16.0 +85 49	12.2	E		102	
3809	N2336	07 18.0 +80 16	11.3	SC			
3838		07 22.3 +72 40	13.9	PEC	MARK		
3841		07 22.6 +30 04	15.2	DBLE			
3847	N2363	07 23.1 +69 17	15.5	IRR			
3851	N2366	07 23.6 +69 18	11.6	IRR	MARK		
3852	I2184	07 23.6 +72 14	13.8	TRPL	MARK		
3857	N2379	07 24.2 +33 54	14.9	S0 COMP		95	3872
3860		07 24.8 +40 51	15.5	IRR			
3872	N2389	07 25.8 +33 57	13.5	SC		95 BR IN GRP	3872

UGC	V_r (km s^{-1})	Spectrum	Disp. (A/mm)	Ref.	Notes
3647					MAG NOT ZW
	1382	N	21-CM	FT	
3653					
	2337	E	401-500	L	
3677	MULT VEL				
	3843	A	401-500	MW	
	3951	A	401-500	L	
3690					MAG NOT ZW
	3145	N	21-CM	FT	
3693					
	2361	E	401-500	MW	
3695					
	5760	A	101-200	PT	
3697	MULT VEL				
	3226	E	301-400	V 5	
	3170	E	301-400	BB42	
	3120	N	21-CM	BG 5	
3708	MULT VEL				
	5013	E	301-400	KR	
	5219	E	1-100	PR	
3709	MULT VEL				
	5241	E	301-400	KR	
	5256	E	1-100	PR	
3730	MULT VEL				
	2611	A	201-300	K	
	2516	E	301-400	A 1	
3734	MULT VEL				
	914	A	301-400	V 5	
	962	N	21-CM	BG 3	
3740	MULT VEL				
	2391	A	401-500	L	
	2400	N	21-CM	PS	
3759					
	4521	E	401-500	L	
3798	MULT VEL				
	1946	A	401-500	MW	
	2088	A	401-500	L	
3809	MULT VEL				
	2252	A	401-500	L	
	2216	N	21-CM	BG 3	
3838	MULT VEL				
	3045	E	101-200	SA 1	
	3163	E	301-400	BG 6	
	3085	E	NO INFO	DS 2	
	3090	N	21-CM	BG 6	
3841					
	5730	E	NO INFO	UL17	
3847					MAG NOT ZW
	17	E	301-400	V 8	
3851	MULT VEL				
	194	E	401-500	L	
	145	E	301-400	V 5	
	102	N	21-CM	FT	
	96	N	21-CM	DD	
	90	N	21-CM	GU	
3852	MULT VEL				
	3420	E	101-200	WK 1	
	3630	E	301-400	BG 6	
	3577	E	301-400	CA	
	3240	E	101-200	KH 1	
	3570	N	21-CM	BG 6	
3857					
	4030	A	401-500	MW	
3860					
	355	N	21-CM	FT	
3872					
	3816	E	401-500	L	

UGC	NGC	R.A. (1950) DEC	m_{pg}	Hubble Type	Char.	Cluster	Group
3918	N2403	07 32.0 +65 43	9.3	SC			5318
3930	N2415	07 33.6 +35 21	12.5	PEC			
3933		07 34.0 +42 03	14.5	SB-C			
3949		07 36.0 +48 51	14.9	SC		103	
3966		07 38.0 +40 13	16.0	IRR		106	
3973		07 38.8 +49 55	13.3	SB	SEYF	103	
3974		07 39.0 +16 55	15.4	IRR			
3984		07 39.9 +70 10	14.2	SB SC			
3995		07 41.0 +29 21	13.6	SB	DISR		
4013		07 43.2 +61 03	14.0	SB	SEYF	100	
4016	N2444	07 43.5 +39 08	13.1	DBLE E	DISR	105	
4017	N2445	07 43.5 +39 08	13.1	DBLE	DISR	105	
4020		07 43.8 +59 08	14.7	SB SC		100	4020 BR IN GRP
4028		07 44.8 +74 28	12.7	S...	MARK		
4030		07 45.1 +28 20	14.4	DBLE		108	
4036	N2441	07 46.1 +73 08	12.7	SC			
4045		07 46.9 +34 33	15.5	COMP			
4047		07 47.0 +30 51	14.5	SB	DISR	108	
4079		07 51.0 +55 50	13.6	S...	MARK	112	
4093	I2209	07 52.0 +60 26	14.5	S...	MARK	100	
4097	N2460	07 52.6 +60 30	12.5	SB		100	
4114	N2475	07 54.1 +53 00	13.9	DBLE E			
4121		07 54.8 +58 12	16.0	SC-I			
4125	N2484	07 55.1 +37 55	14.9	SO			

UGC	V_r (km s^{-1})	Spectrum	Disp. (A/mm)	Ref.	Notes
3918	MULT VEL				
	119	E	201-300	L S	
	70	E	401-500	M W	
	125	E	1-100	DH 2	
	128	N	21-CM	R R	
	133	N	21-CM	GW 1	
	138	N	21-CM	R B	
3930	MULT VEL				
	3822	E	301-400	L B	
	3786	N	21-CM	BG 2	
3933					
	5856	E	101-200	FR11	
3949					
	6358	E	101-200	FR11	
3966					MAG NOT ZW
	364	N	21-CM	FT	
3973	MULT VEL				
	6597	E	1-100	SA 3	
	6600	E	201-300	AD 1	
	6637	E	201-300	UL10	
	5201	N	NO INFO	SA 2	
	6580	E	101-200	WK 2	
					DISCREPANT VELOCITIES?
3974					
	265	N	21-CM	FT	
3984					
	3894	E	101-200	FR11	
3995					
	4671	E	301-400	KR	
4013					
	8700	E	101-200	WK 1	
4016					MAG SHARED
	3965	A	301-400	BB 2	
4017					MAG SHARED
	4000	E	301-400	BB 2	
4020					
	6508	E	101-200	FR11	
4028	MULT VEL				
	4003	E	101-200	SA 1	
	4025	E	301-400	BG 6	
	4070	E	NO INFO	DS 2	
	4130	N	21-CM	BG 4	
4030	MULT VEL				
	8450	E	301-400	KR	COMP VEL A
	8075	E	301-400	KR	COMP VEL B
4036					
	3623	A	401-500	L	
4045					
	8520	A	201-300	UL15	
4047					
	4363	N	101-200	FR11	
4079	MULT VEL				
	6055	E	101-200	SA 3	
	6300	E	201-300	AD 2	
4093					
	1560	E	101-200	WK 1	
4097					
	1442	E	401-500	MW	
4114					
	5019	A	401-500	L	
4121	MULT VEL				MAG NOT ZW
	1090	N	21-CM	FT	
	1104	N	21-CM	BA 4	
4125	MULT VEL				4C37.21
	12990	E	301-400	BU	
	12390	E	NO INFO	UL17	
					DISCREPANT VELOCITIES?

UGC	NGC	R.A. (1950) DEC	m pg	Hubble Type	Char.	Cluster	Group
4139		07 56.5 +16 34	15.0	SC			4172
4165	N2500	07 58.2 +50 54	12.3	... SC			
4189	N2514	08 00.0 +15 57	14.4	SC			4172
4191	N2512	08 00.1 +23 32	14.2	SB	MARK		
4200		08 01.1 +40 21	14.6	SB SC		114	
4220	I2226	08 03.4 +12 41	14.9	SB			
4229		08 04.2 +39 09	14.4	...	MARK	114	
4231	N2526	08 04.3 +08 09	14.6	...			
4242		08 05.4 +72 57	14.4	... E-S0	MARK		
4256	N2532	08 07.0 +34 06	12.9	SC			
4260		08 07.6 +46 37	14.3	IRR			
4264	N2535	08 08.2 +25 20	13.5	SC SB			
4268	N2534	08 09.0 +55 49	13.8	PEC	MARK	121	
4271	N2523	08 09.3 +73 44	12.4	SB			
4274	N2537	08 09.7 +46 09	11.7	S... SC	MARK		
4284	N2541	08 11.0 +49 13	13.0	SC			
4287	N2545	08 11.4 +21 30	13.2	S... SC		118	
4289		08 11.7 +58 29	14.6	E		116	
4305		08 14.1 +70 52	11.3	IRR			
4312	N2554	08 14.9 +23 37	13.5	S0-A		118	
4313	N2549	08 14.9 +57 58	12.1	S0		116	
4324		08 15.7 +20 55	15.3	S...		118	
4325	N2552	08 15.7 +50 10	13.5	IRR SC-I			

UGC	V_r (km s^{-1})	Spectrum	Disp. (A/mm)	Ref.	Notes
4139					
	4848	E	101-200	FR11	
4165	MULT VEL				
	470	E	401-500	L	
	530	N	21-CM	GU	
4189					
	4913	N	101-200	FR11	
4191					
	4800	E	201-300	AD 5	
4200					
	12202	N	101-200	FR11	
4220					
	10906	E	101-200	FR11	
4229					
	7075	E	101-200	DS 3	
4231					
	4634	E	1-100	PR	
4242					
	3150	E	401-500	WK 1	
4256					
	5153	E	401-500	MW	
4260					
	2254	N	21-CM	FT	
4264	MULT VEL				
	4056	E	101-200	FR 8	
	4243	E	401-500	MW	
	4050	E	301-400	P	
	4110	E	1-100	A 4	
	4109	N	21-CM	FR 8	
	4110	N	21-CM	PS	
	4101	E	1-100	PR	
4268					
	3517	A	101-200	SA 3	
4271					
	3448	A	401-500	L	
4274	MULT VEL				
	397	E	201-300	MW	
	290	E	301-400	L	
	404	E	301-400	LB	
	450	E	201-300	AD 4	
	441	E	NO INFO	DS 2	
4284	MULT VEL				
	601	E	301-400	V 5	
	800	N	301-400	V 8	
	570	N	21-CM	GU	
	561	N	21-CM	S 3	
4287	MULT VEL				
	3192	E	201-300	CR 2	
	3501	A	101-200	SN 8	
4289					
	7890	A	101-200	PT	
4305	MULT VEL				
	220	E	401-500	L	
	158	N	21-CM	FT	
	156	N	21-CM	DD	
	150	N	21-CM	RR	
4312					
	4163	A	201-300	CR 2	
4313					
	1082	A	401-500	MW	
4324					
	4801	A	201-300	K	
4325	MULT VEL				
	974	A	301-400	V 8	
	511	N	21-CM	BA 1	
	478	E	101-200	SN 8	

UGC	NGC	R.A. (1950) DEC			m pg	Hubble Type	Char.	Cluster	Group
4327	N2544	08 15.9	+74	08	13.4	SA	MARK		
4330	N2557	08 16.3	+21	36	14.6	S0		118	
4332		08 16.8	+21	16	15.5	PEC		118	
4334	N2565	08 16.9	+22	11	13.8	SB		118	
4337	N2560	08 17.0	+21	08	14.9	S0-A		118	
4344		08 17.4	+21	02	15.5	...		118	
4345	N2562	08 17.5	+21	17	14.0	S0-A SA		118	
4347	N2563	08 17.7	+21	14	13.7	S0		118	
4362	N2551	08 19.1	+73	35	12.7	S... SA-B			
4363		08 19.3	+74	36	14.9	SC	DISR		
4367	N2577	08 19.8	+22	43	13.8	E-S0		118	
4383	I2338	08 20.7	+21	30	14.7	DBLE S... S...		118	
4384	I2341	08 20.8	+21	36	14.9	E-S0		118	
4391	N2582	08 22.4	+20	30	14.3	SB		118	
4399		08 23.2	+21	37	15.5	SC-I		118	
4417		08 24.3	+55	52	14.3	COMP	MARK	121	
4422	N2595	08 24.8	+21	39	13.9	SB-C		118	
4426		08 25.2	+42	02	18.0	IRR		127	
4438		08 26.2	+52	52	13.9	S...	MARK	124	
4442		08 27.0	+52	28	14.5	SB-C SC		124	
4456	I 509	08 29.0	+24	10	14.6	SC			
4457		08 29.1	+19	23	14.9	S...			
4458	N2599	08 29.2	+22	44	13.4	SA	MARK	118	
4484	N2608	08 32.2	+28	39	13.2	SB-C SA			
4490		08 33.1	+66	20	16.0	S...	MARK	129	

UGC	V_r (km s^{-1})	Spectrum	Disp. (A/mm)	Ref.	Notes
4327	MULT VEL				
	2736	E	101-200	SA 3	
	2851	E	NO INFO	DS 1	
4330					
	4944	A	201-300	K	
4332					
	5505	E	201-300	TI 3	
4334	MULT VEL				
	3422	A	201-300	CR 2	
	3599	A	201-300	TI 3	
	3684	N	NO INFO	HG	
4337					
	4903	A	201-300	K	
4344					
	5008	N	101-200	SA 4	
4345					
	4963	A	401-500	MW	
4347	MULT VEL				
	4775	A	401-500	MW	
	4391	A	201-300	TI 3	
4362					
	2296	A	401-500	L	
4363					
	3523	N	21-CM	FT	
4367					
	2123	A	201-300	K	
4383	MULT VEL				
	5230	E	201-300	K	COMP VEL A
	5194	E	201-300	K	COMP VEL B
	5400	N	21-CM	PS	
4384					
	4846	A	201-300	CR 2	
4391	MULT VEL				
	4383	A	201-300	CR 2	
	4514	N	101-200	FR11	
4399					
	4349	N	101-200	SA 4	
4417	MULT VEL				
	9139	E	101-200	SA 3	
	9161	E	301-400	SA 2	
	9300	E	201-300	AD 2	
	9175	E	201-300	UL15	
4422	MULT VEL				
	4111	A	301-400	SA 2	
	4367	E	201-300	CR 2	
	4585	A	201-300	BR 2	
4426					MAG NOT ZW
	392	N	21-CM	FT	
4438	MULT VEL				
	4231	E	101-200	SA 3	
	4200	E	201-300	AD 4	
	4308	E	NO INFO	DS 1	
4442					
	5090	N	101-200	FR11	
4456					
	5494	E	101-200	FR11	
4457					
	11190	E	NO INFO	A 4	
4458					
	4690	E	201-300	CR 2	
4484	MULT VEL				
	2119	A	401-500	MW	
	2126	N	21-CM	BI 1	
4490	MULT VEL				MAG NOT ZW
	5226	E	101-200	SA 3	
	4800	E	201-300	AD 1	
	5288	E	NO INFO	DS 2	

UGC	NGC	R.A. (1950) DEC	m_{pg}	Hubble Type	Char.	Cluster	Group
4499		08 34.0 +51 50	14.0	...			
4508		08 35.4 -02 17	14.5	SC-I COMP		126	
4509	N2623	08 35.4 +25 56	14.4	TRPL SC		128	
4544	N2639	08 40.0 +50 22	12.4	SA			
4572		08 42.4 +37 07	13.8	COMP	MARK		
4574	N2633	08 42.6 +74 17	12.4	SB			
4576	I2389	08 42.7 +73 43	13.2	S... S0	DISR		
4587		08 43.7 +49 44	13.8	S0			
4604	N2646	08 45.1 +73 39	13.0	S0			
4605	N2654	08 45.2 +60 25	12.8	SA-B SA			
4619	N2672	08 46.5 +19 16	13.4	E		131	
4620	N2673	08 46.5 +19 16	14.4	... E		131	
4627	N2676	08 48.0 +47 44	14.3	S0			
4637	N2655	08 49.0 +78 25	10.8	S0-A			
4641	N2683	08 49.6 +33 36	9.7	SB			
4645	N2681	08 49.9 +51 31	10.4	SA		135	
4658	I2421	08 51.2 +32 52	14.9	SC			
4662	N2684	08 51.3 +49 20	13.4	S... PEC		132	
4664	N2691	08 51.5 +39 44	13.9	SA	SEYF	127	
4666	N2685	08 51.7 +58 55	12.1	PEC S0			
4671		08 53.1 +52 18	13.6	S...	DISR	135	
4674	N2693	08 53.3 +51 33	13.1	E		135	

UGC	V_r (km s^{-1})	Spectrum	Disp. (A/mm)	Ref.	Notes
4499					
	660	E	201-300	A 9	
4508					
	1940	E	201-300	UL15	
4509	MULT VEL				
	5435	E	401-500	MW	
	5400	N	21-CM	BI 1	
4544					
	3314	A	401-500	MW	
4572					
	3934	E	101-200	DS 3	
4574					
	2228	E	401-500	L	
4576					
	2632	E	401-500	L	
4587					
	3060	N	NO INFO	RT	
4604					
	3546	A	401-500	L	
4605					
	1360	A	401-500	MW	
4619	MULT VEL				
	4223	A	401-500	MW	
	3964	N	21-CM	BI 1	VEL CONTAM
4620	MULT VEL				
	3792	A	401-500	MW	
	3964	N	21-CM	BI 1	VEL CONTAM
4627					
	6010	N	NO INFO	RT	
4637	MULT VEL				
	1299	E	401-500	MW	
	1517	E	NO INFO	BB38	
	1389	N	21-CM	DL 2	
	1450	N	21-CM	KG 1	
4641	MULT VEL				
	310	E	101-200	FR 8	
	336	E	401-500	MW	
	335	E	401-500	L	
	285	E	301-400	LB	
	260	N	21-CM	R 3	
4645	MULT VEL				
	703	A	201-300	MW	
	736	A	401-500	L	
4658					
	4395	E	101-200	FR11	
4662					
	3333	N	NO INFO	HJ	
4664	MULT VEL				
	3900	E	201-300	AD 5	
	3936	E	301-400	BG 6	
	3990	N	21-CM	BG 6	
4666	MULT VEL				
	884	E	201-300	MW	
	867	E	301-400	LB	
	830	E	NO INFO	BB38	
	910	E	201-300	UL 1	
	883	N	21-CM	BA 2	
	875	N	21-CM	PS	
	930	N	21-CM	KG 1	
	870	N	21-CM	BI 2	
4671	MULT VEL				
	4266	E	301-400	KR	
	3661	N	101-200	TU	
					DISCREPANT VELOCITIES?
4674					
	4956	A	401-500	MW	

UGC	NGC	R.A. (1950) DEC	m_{pg}	Hubble Type	Char.	Cluster	Group
4675	N2692	08 53.3 +52 16	14.1	SA-B		135	
4691	N2713	08 54.7 +03 06	12.9	SB			
4692	N2716	08 54.9 +03 16	13.7	SO SA			
4695	N2701	08 55.3 +53 57	12.3	SC		135	
4703		08 55.8 +06 31	15.5	SB-C DBLE			
4707	N2718	08 56.2 +06 30	13.3	SA-B			
4708	N2712	08 56.2 +45 06	12.3	SB		136	
4718	N2719	08 57.1 +35 55	13.7	PEC SA-B			
4723	N2723	08 57.6 +03 22	14.5	SO			
4730		08 58.0 +60 21	14.3	...	MARK		
4744	N2735	08 59.7 +26 08	14.2	...			
4749		09 01.0 +51 50	13.6	S...	MARK	135	
4750	N2726	09 01.0 +60 08	13.1	SA	MARK		
4757	N2744	09 01.8 +18 39	13.7	DBLE SB	DISR	140	4763
4759	N2715	09 01.9 +78 17	11.9	SC			
4763	N2749	09 02.5 +18 30	13.3	E		140 BR IN GRP	4763
4772	N2752	09 02.9 +18 32	14.8	SB		140	4763
4779	N2742	09 03.6 +60 40	12.0	SC			
4794	N2764	09 05.4 +21 39	13.9	S...			
4797		09 05.5 +06 08	15.7	PEC SC-I			
4818	N2732	09 07.1 +79 24	12.6	SO			
4820	N2775	09 07.7 +07 15	11.4	SA			
4821	N2768	09 07.7 +60 15	11.1	E-SO			
4825	N2748	09 08.0 +76 41	11.7	SC			
4829		09 08.4 +46 51	14.3	COMP	MARK		
4837		09 09.0 +35 44	15.5	SC-I			
4838	N2776	09 09.0 +45 10	12.1	SC			
4840	N2778	09 09.2 +35 13	13.1	E		BR IN GRP	4840
4843	N2780	09 09.6 +35 07	14.2	S...			4840

UGC	V_r (km s^{-1})	Spectrum	Disp. (A/mm)	Ref.	Notes
4675					
	3749	N	101-200	TU	
4691	MULT VEL				
	3939	A	201-300	SN 8	
	3810	A	101-200	FR 4	
4692					
	3537	A	401-500	MW	
4695	MULT VEL				
	2338	N	21-CM	BG 1	
	2299	A	101-200	SN 8	
4703					
	3556	E	NO INFO	Z 3	
4707					
	3848	N	21-CM	CX	
4708	MULT VEL				
	1840	A	401-500	MW	
	1832	N	NO INFO	HJ	
4718					MAG SHARED
	3190	A	301-400	P	
4723					
	3725	A	401-500	MW	
4730					
	3250	E	NO INFO	DS 1	
4744					
	2670	A	NO INFO	Z 3	
4749	MULT VEL				
	4758	E	101-200	SA 3	
	4750	N	21-CM	BG 4	
	4800	E	NO INFO	DS 1	
4750					
	1401	A	101-200	SA 1	
4757					
	3450	E	401-500	MW	
4759					
	1158	E	401-500	L	
4763	MULT VEL				
	4203	E	401-500	MW	
	4270	E	NO INFO	BB38	
4772					
	4022	A	201-300	CR 2	
4779					
	1291	E	301-400	V 5	
4794					
	2597	E	201-300	SN 8	
4797					
	1305	N	21-CM	FT	
4818					
	2121	A	401-500	L	
4820					
	1135	A	401-500	MW	
4821					
	1408	A	401-500	MW	
4825					
	1489	E	401-500	L	
4829	MULT VEL				
	4239	E	101-200	SA 3	
	4302	E	NO INFO	DS 1	
4837					
	1880	N	21-CM	FT	
4838	MULT VEL				
	2673	A	401-500	L	
	2620	N	21-CM	DU	
4840					
	2051	N	101-200	TU	
4843					
	2208	N	101-200	TU	

56

UGC	NGC	R.A. (1950) DEC	m pg	Hubble Type	Char.	Cluster	Group
4862	N2782	09 10.9 +40 20	12.3	S... SA			
4864		09 11.4 +16 57	14.7	SA SC			
4881		09 12.7 +44 33	14.9	DBLE	DISR		
4894	N2793	09 13.7 +34 39	13.9	S... PEC		142	
4897	N2802	09 13.9 +19 10	14.3	DBLE E		140	
4898	N2803	09 13.9 +19 10	14.3	DBLE E		140	
4905	N2798	09 14.1 +42 12	12.9	SA	DISR	BR IN GRP	4905
4914	N2787	09 14.8 +69 25	11.7	SO SA			
4919		09 15.0 +45 52	14.4	SC			
4933	N2824	09 16.1 +26 29	14.3	SO	MARK		
4935	N2823	09 16.2 +34 13	15.7	SA SO		142	
4936	N2805	09 16.3 +64 19	11.9	SC	DISR	BR IN GRP	4936
4942	N2832	09 16.8 +33 58	13.6	E		142 BR IN CLUS	
4951		09 17.0 +71 45	14.7	...	MARK		
4952	N2814	09 17.1 +64 28	14.0	IRR			4936
4961	N2820	09 17.8 +64 29	13.1	SC			4936
4966	N2841	09 18.5 +51 12	9.9	SB			
4971	N2844	09 18.6 +40 22	13.6	SO-A SA		145 BR IN GRP	4971
4986	N2852	09 20.0 +40 23	14.0	COMP SO-A		145	4971
4987	N2853	09 20.1 +40 25	14.6	SA		145	4971
5000	N2857	09 21.1 +49 34	14.3	SC		143	4995
5001	N2859	09 21.3 +34 44	11.8	SO			

UGC	V_r (km s^{-1})	Spectrum	Disp. (A/mm)	Ref.	Notes
4862	MULT VEL				
	2517	E	201-300	MW	
	2358	E	301-400	L B	
	2526	N	NO INFO	D A 1	
	2538	N	301-400	D A 2	
	2536	E	101-200	S O 1	
	2561	N	21-CM	B A 2	
	2554	E	1-100	K T 5	
4864					
	8263	N	101-200	FR11	
4881					
	11773	E	301-400	A 1	
4894					
	1669	E	301-400	V 5	
4897					MAG SHARED
	8634	N	101-200	T U	
4898					MAG SHARED
	8832	N	101-200	T U	
4905					
	1708	E	401-500	MW	
4914	MULT VEL				
	639	A	401-500	MW	
	551	A	401-500	L	
4919					
	8096	E	101-200	FR11	
4933					
	9300	E	201-300	A D 5	
4935	MULT VEL				
	6960	A	NO INFO	UL17	
	5035	A	101-200	SN 8	
					DISCREPANT VELOCITIES?
4936	MULT VEL				
	1688	E	101-200	FR 8	
	1916	E	401-500	L	
	1742	N	21-CM	FR 8	
4942	MULT VEL				
	6946	A	401-500	MW	
	6803	N	101-200	T U	
	6934	A	101-200	SN 8	
4951	MULT VEL				
	3491	E	101-200	SA 1	
	3600	E	201-300	AD 4	
	3601	E	NO INFO	DS 2	
4952					
	1662	E	301-400	P	
4961	MULT VEL				
	1702	E	301-400	P	COMP VEL A
	1667	E	301-400	P	COMP VEL B
4966	MULT VEL				
	584	E	201-300	MW	
	740	E	401-500	L	
	600	N	NO INFO	SB	
	641	N	21-CM	BG 3	
	661	N	21-CM	R 3	
4971					
	1455	E	101-200	SN 8	
4986					
	1888	N	101-200	T U	
4987					
	1800	N	101-200	T U	
5000					
	4864	E	101-200	FR11	
5001	MULT VEL				
	1694	A	401-500	MW	
	1587	N	21-CM	B I 1	

UGC	NGC	R.A. (1950) DEC	m_{pg}	Hubble Type	Char.	Cluster	Group
5018	N2872	09 23.0 +11 39	13.0	E			
5021	N2874	09 23.1 +11 39	13.5	SC			
5023		09 23.2 +19 36	14.4	...	MARK		
5028		09 23.6 +68 39	13.9	PEC	MARK		
5029		09 23.8 +68 40	14.3	SC	DISR		
5043	I2476	09 24.9 +30 13	14.5	E-S0		148	5060
5051	N2880	09 25.7 +62 43	12.6	S0			
5055		09 26.6 +56 04	14.5	SB	MARK		
5060	N2893	09 27.3 +29 46	13.6	SA	MARK	148 BR IN GRP	5060
5063		09 27.4 +49 28	15.4	COMP PEC	MARK		
5079	N2903	09 29.4 +21 44	9.8	SB-C SC			
5092	N2911	09 31.1 +10 22	13.6	... S0		BR IN GRP	5092
5096	N2914	09 31.4 +10 20	13.7	SA			5092
5130	N2936	09 35.1 +02 58	14.4	MULT E	DISR		
5131	N2937	09 35.1 +02 58	14.4	MULT E			
5134	N2939	09 35.4 +09 45	13.5	SB-C		BR IN GRP	5134
5139		09 36.0 +71 25	15.5	IRR			5318
5140	N2942	09 36.1 +34 14	14.1	SC			
5144	N2944	09 36.3 +32 33	14.7	DBLE	DISR		

UGC	V_r (km s^{-1})	Spectrum	Disp. (A/mm)	Ref.	Notes
5018	MULT VEL				
	2954	A	301-400	P	
	3710	N	21-CM	BI 1	VEL CONTAM
					DISCREPANT VELOCITIES?
5021	MULT VEL				
	3591	A	301-400	P	
	3710	N	21-CM	BI 1	VEL CONTAM
5023					
	2400	E	201-300	AD 5	
5028	MULT VEL				
	3638	E	101-200	SA 3	
	3924	E	301-400	KR	
	3600	E	201-300	AD 1	
5029					
	3860	E	301-400	KR	
5043					
	7980	A	NO INFO	UL17	
5051					
	1514	A	401-500	MW	
5055	MULT VEL				
	7470	E	101-200	SA 3	
	7537	E	NO INFO	DS 1	
	6000	E	201-300	AD 4	
					DISCREPANT VELOCITIES?
5060	MULT VEL				
	1800	E	201-300	AD 5	
	1680	E	301-400	BG 6	
	1711	N	21-CM	BG 6	
5063	MULT VEL				
	7758	E	101-200	SA 3	
	7800	E	201-300	AD 2	
5079	MULT VEL				
	585	E	301-400	BB 9	
	642	E	401-500	MW	
	645	E	401-500	L	
	540	E	1-100	WD 1	
	580	N	21-CM	GU	
	561	N	21-CM	R 3	
5092	MULT VEL				
	3032	E	1-100	DC	
	3140	E	401-500	MW	
	3225	E	NO INFO	BB38	
5096					
	3370	A	401-500	MW	
5130	MULT VEL				MAG SHARED
	7039	A	301-400	A 1	
	7393	E	201-300	CR 2	
	6867	N	NO INFO	FV	
5131	MULT VEL				MAG SHARED
	6927	A	201-300	CR 2	
	6899	A	301-400	A 1	
	7167	N	NO INFO	FV	
	6705	A	201-300	K	
5134					
	3367	N	101-200	TU	
5139	MULT VEL				
	141	N	21-CM	FT	
	140	N	21-CM	BG 9	
	155	N	21-CM	DL 2	
5140	MULT VEL				
	4478	N	101-200	FR11	
	4721	A	101-200	SN 8	
5144	MULT VEL				
	6983	E	301-400	CA	
	6831	N	21-CM	CX	

UGC	NGC	R.A. (1950) DEC	m pg	Hubble Type	Char.	Cluster	Group
5146		09 36.4 +32 36	14.3	DBLE S... SA	DISR		
5166	N2955	09 38.2 +36 07	13.9	SB SC			
5167	N2962	09 38.3 +05 24	13.1	S0-A			
5176	N2950	09 39.0 +59 05	11.8	S0 S0			
5180	N2967	09 39.5 +00 34	12.2	SC			
5183	N2964	09 39.9 +32 05	12.0	SB-C	MARK		5183 BR IN GRP
5188	N2909	09 40.0 +66 13	14.1	PEC	MARK		
5190	N2968	09 40.2 +32 10	13.1	IRR SB			5183
5221	N2976	09 43.1 +68 09	10.9	SC			
5222	N2963	09 43.2 +73 12	14.3	SA-B	MARK		
5225		09 43.3 +46 00	14.9	COMP			
5229	N2990	09 43.6 +05 56	12.5	S... SC			
5230	I 564	09 43.8 +03 18	14.1	IRR E-S0			
5243		09 44.9 +58 13	15.1	SB	MARK		
5250	N2998	09 45.5 +44 18	13.3	SC			5250 BR IN GRP
5251	N3003	09 45.6 +33 39	12.3	SC SB			
5253	N2985	09 45.8 +72 31	11.1	SB			
5259	N3011	09 46.7 +32 27	14.2	S0	MARK		
5266	N3016	09 47.1 +12 56	13.7	SB			5271
5271	N3020	09 47.4 +13 03	13.2	SC			5271 BR IN GRP
5272		09 47.4 +31 43	14.7	IRR			
5275	N3024	09 47.8 +13 00	13.7	S...			5271
5280	N3021	09 48.0 +33 47	12.6	S... SB-C			
5292	N3032	09 49.2 +29 28	13.0	SC S0			
5295		09 49.7 +43 05	14.5	SB SC			5295 BR IN GRP
5303	N3041	09 50.4 +16 55	13.1	SC			

UGC	V_r (km s^{-1})	Spectrum	Disp.	Ref.	Notes
			(A/mm)		
5146	MULT VEL				
	6580	E	201-300	K	COMP VEL A
	6377	A	201-300	K	COMP VEL B
	6470	A	201-300	BR 2	
5166	MULT VEL				
	1793	A	301-400	V 8	
	7026	E	101-200	SN 8	
					DISCREPANT VELOCITIES?
5167	MULT VEL				
	1970	A	201-300	CR 2	
	2116	N	21-CM	BI 1	
	2009	A	101-200	SN 8	
5176	MULT VEL				
	1430	A	401-500	MW	
	1339	A	401-500	L	
5180					
	2245	A	401-500	L	
5183	MULT VEL				
	1080	E	101-200	DS 3	
	1340	E	401-500	MW	
	1200	E	201-300	AD 7	
5188					
	3000	E	201-300	AD 2	
5190	MULT VEL				
	1608	A	301-400	V 5	
	1570	A	101-200	SN 8	
5221					
	42	E	401-500	L	
5222					
	6533	E	NO INFO	DS 1	
5225					
	4936	A	301-400	SA 2	
5229					
	3087	E	201-300	SN 8	
5230					
	6093	E	301-400	P	
5243	MULT VEL				
	8504	E	101-200	SA 1	
	8416	E	NO INFO	DS 2	
5250					
	4690	A	101-200	SN 8	
5251					
	1476	E	401-500	MW	
5253					
	1277	A	401-500	MW	
5259	MULT VEL				
	1290	E	101-200	DS 3	
	1500	E	201-300	AD 7	
5266					
	8844	N	101-200	KI	
5271					
	1429	N	101-200	KI	
5272	MULT VEL				
	520	N	21-CM	FT	
	380	N	101-200	BF	COMP VEL A
	373	N	101-200	BF	COMP VEL B
5275					
	1506	N	101-200	KI	
5280	MULT VEL				
	1529	E	301-400	V 5	
	1465	E	201-300	SN 8	
5292	MULT VEL				
	1568	A	401-500	MW	
	1501	N	21-CM	BI 1	
5295					
	4786	A	101-200	SN 8	
5303	MULT VEL				
	1417	N	21-CM	BG 1	
	1313	A	101-200	SN 8	

UGC	NGC	R.A. (1950) DEC	m_{pg}	Hubble Type	Char.	Cluster	Group
5311	N3044	09 51.0 +01 48	12.4	SC			
5316	N3027	09 51.2 +72 27	12.3	SC			
5318	N3031	09 51.4 +69 18	8.1	SB			5318 BR IN GRP
5322	N3034	09 51.7 +69 55	9.2	IRR	DISR		5318
5327	N3043	09 52.6 +59 32	13.3	S...			
5328	N3055	09 52.7 +04 31	12.3	SC			
5340		09 53.8 +29 04	15.2	PEC			
5351	N3067	09 55.4 +32 37	12.7	SA-B			
5353	N3068	09 55.7 +29 07	15.1	E-SO	DISR		
5364		09 56.4 +30 59	14.6	SO IRR			99999
5366	N3074	09 56.7 +35 38	14.8	SC			
5373		09 57.4 +05 34	12.2	IRR			
5375	N3065	09 57.6 +72 25	12.9	SO			
5379	N3066	09 57.9 +72 22	12.8	S...	MARK		
5387	N3079	09 58.5 +55 55	11.2	SB ...		163	5387 BR IN GRP
5397	N3098	09 59.4 +24 57	13.0	SC SO-A			
5398	N3077	09 59.4 +68 58	10.7	SO IRR			5318
5404	N3057	10 00.0 +80 32	14.2	... SC-I			
5408		10 00.3 +59 40	14.2	COMP	MARK		

UGC	V_r (km s^{-1})	Spectrum	Disp. (A/mm)	Ref.		Notes
5311	MULT VEL					
	1326	E	301-400	V	5	
	1359	E	1-100	V	6	
5316	MULT VEL					
	1079	E	401-500	L		:NCE NOTES
	1061	N	21-CM	BG	3	
5318	MULT VEL					
	-42	E	1-100	FR	8	
	-55	E	201-300	MW		
	-64	E	401-500	L		
	-30	N	NO INFO	SB		
	-45	N	NO INFO	MU		
	-46	E	1-100	GO		
	-40	N	21-CM	FR	8	
5322	MULT VEL					3C231
	263	A	201-300	MW		
	275	A	401-500	L		
	305	E	301-400	BB	33	
	379	E	1-100	HK		
	290	N	NO INFO	SB		
	281	N	201-300	DA	2	
	239	E	101-200	FR	5	
	256	A	1-100	V	6	
	240	N	21-CM	WE	2	
5327						
	2905	E	201-300	SN	8	
5328						
	1913	E	401-500	L		
5340						
	504	N	21-CM	FT		
5351	MULT VEL					
	1506	E	401-500	MW		
	1399	E	101-200	DN		
5353						MAG SHARED
	6409	E	101-200	ZH	1	
5364						
	26	N	21-CM	FT		
5366	MULT VEL					
	5161	E	101-200	FR	11	
	5000	E	101-200	KZ		
5373						
	295	N	21-CM	FT		
5375	MULT VEL					
	2051	E	401-500	L		
	1984	E	NO INFO	BB	38	
5379	MULT VEL					
	2033	E	201-300	UL	10	
	1800	E	201-300	AD	2	
	2132	E	401-500	L		
	2070	E	1-100	WK	3	
5387	MULT VEL					
	1168	E	301-400	LB		
	1171	E	401-500	L		
	1150	E	1-100	CA		
5397						
	1310	A	101-200	SN	8	
5398	MULT VEL					
	-158	E	401-500	L		
	-41	E	101-200	UL	7	
	-10	E	1-100	BT		
	10	N	21-CM	DL	2	
5404						
	1529	N	21-CM	FT		
5408	MULT VEL					
	2730	E	401-500	WK	1	
	3000	E	1-100	WD	1	

UGC	NGC	R.A. (1950) DEC	m pg	Hubble Type	Char.	Cluster	Group
5414	N3104	10 00.9 +41 00	14.2	IRR			
5425		10 01.7 +13 52	13.6	... SC			
5443	I 590	10 03.3 +00 53	14.2	DBLE E E		158	
5478		10 06.6 +30 24	15.5	IRR			
5491		10 08.6 +59 07	15.5	SB	MARK		
5494		10 09.0 +67 40	16.0	PEC			
5503	N3156	10 10.1 +03 22	12.8	SO E			5516
5510	N3162	10 10.8 +22 59	12.2	SC		164	
5511	N3158	10 10.8 +39 01	13.4	E		161 BR IN GRP	5511
5516	N3166	10 11.1 +03 40	11.1	SO-A SA		BR IN GRP	5516
5517	N3163	10 11.1 +38 55	14.4	E-SO E		161	5511
5525	N3169	10 11.7 +03 43	11.9	SA	DISR		5516
5532	N3147	10 12.7 +73 38	11.3	SB-C SB			
5536	N3168	10 13.0 +60 29	14.6	E COMP		BR IN GRP	5536
5543		10 13.7 +05 04	14.6	SC			
5544	N3177	10 13.8 +21 22	12.8	SB		164	5559
5549		10 14.0 +53 43	15.3	SB SC		163	
5554	N3185	10 14.9 +21 56	12.9	SA		164	5559
5556	N3187	10 15.1 +22 07	13.8	S... SC-I	DISR	164	5559
5557	N3184	10 15.3 +41 40	10.4	SC			
5559	N3190	10 15.4 +22 05	11.9	SA		164 BR IN GRP	5559
5561	I 602	10 15.7 +07 18	13.4	S...			
5562	N3193	10 15.7 +22 09	12.4	E		164	5559
5565	N3191	10 16.0 +46 43	13.9	S... SC			
5569	N3188	10 16.4 +57 40	14.7	SB	MARK		
5572	N3198	10 16.9 +45 49	10.7	SC			
5581	N3202	10 17.5 +43 16	14.2	SA SC		BR IN GRP	5581
5589	N3206	10 18.5 +57 12	12.7	SC		180	

UGC	V_r (km s^{-1})	Spectrum	Disp. (A/mm)	Ref.	Notes
5414	MULT VEL				
	725	E	301-400	V 5	
	630	N	21-CM	BA 1	
5425					
	2713	E	101-200	SN 8	
5443	MULT VEL				
	6216	N	101-200	TU	COMP VEL A
	6383	N	101-200	TU	COMP VEL B
5478					
	1371	N	21-CM	FT	
5491					
	9132	E	NO INFO	DS 1	
5494					MAG NOT ZW
	4500	E	201-300	AD 4	
5503					
	1144	A	201-300	SN 8	
5510					
	1456	A	401-500	MW	
5511					
	7024	A	401-500	MW	
5516					
	1381	E	401-500	MW	
5517					
	6245	A	401-500	L	
5525	MULT VEL				
	1281	E	401-500	MW	
	1312	E	401-500	L	
	1093	N	21-CM	BG 3	
5532					
	2721	A	401-500	MW	
5536					
	9283	N	101-200	TU	
5543					
	13756	A	101-200	SN 8	
5544					
	1220	A	401-500	MW	
5549					
	13592	A	101-200	SN 8	
5554					
	1241	E	401-500	MW	
5556					
	1589	E	301-400	V 5	
5557	MULT VEL				
	395	A	401-500	L	
	443	E	401-500	MW	
	588	N	21-CM	DL 2	
	595	N	21-CM	S 3	
5559	MULT VEL				
	1380	A	401-500	L	
	1319	E	201-300	MW	
5561					
	3808	N	101-200	TU	
5562					
	1371	A	401-500	MW	
5565					
	9115	E	101-200	SN 8	
5569	MULT VEL				
	7748	E	101-200	SA 1	
	7788	E	NO INFO	DS 2	
5572	MULT VEL				
	649	E	401-500	L	
	645	N	21-CM	GU	
	677	N	21-CM	R 3	
5581					
	6715	A	101-200	SN 8	
5589	MULT VEL				
	1192	E	301-400	V 5	
	1158	N	21-CM	S 3	

UGC	NGC	R.A. (1950) DEC	m pg	Hubble Type	Char.	Cluster	Group
5604		10 19.7 +46 30	14.8	SC			
5610	N3222	10 19.9 +20 07	14.5	E-SO SO		169	
5612		10 20.1 +71 08	14.8	... SC-I		184	
5615		10 20.6 +53 21	14.2	DBLE S...		163	
5617	N3226	10 20.7 +20 08	13.3	S... E		169	5559
5620	N3227	10 20.8 +20 06	12.2	SB	SEYF	169	5559
5633		10 22.0 +15 00	15.4	... SC-I			
5637	N3239	10 22.4 +17 25	13.5	PEC IRR		173	
5663	N3245	10 24.5 +28 46	11.6	SO			
5666	I2574	10 24.8 +68 40	11.2	SC-I IRR			
5674	N3253	10 25.8 +12 57	14.4	SB-C SC			
5685	N3254	10 26.5 +29 45	12.4	SB SC			
5688		10 26.6 +70 19	14.6	... SC-I		184	
5717	N3259	10 29.0 +65 19	12.9	SB			
5720		10 29.4 +54 38	13.2	PEC	MARK	163	
5721	N3274	10 29.5 +27 56	13.3	IRR S...			
5731	N3277	10 30.1 +28 47	12.3	SA-B SA			
5742	N3287	10 32.0 +21 55	12.9	IRR SC-I		185	
5744		10 32.1 +46 49	14.1	COMP	MARK		
5747		10 32.7 +44 35	15.4	PEC		179	
5752	N3288	10 33.2 +58 49	15.0	SB			
5753	N3294	10 33.3 +37 35	11.5	SC			
5764		10 33.9 +31 49	15.6	IRR			
5766	N3300	10 34.0 +14 26	13.4	SO			

66

JGC	V_r (km s^{-1})	Spectrum	Disp. (A/mm)	Ref.	Notes
604					
	4983	N	101-200	FR11	
610					
	5577	A	201-300	MW	
612					
	1011	N	21-CM	FT	
615	MULT VEL				
	9675	N	101-200	TU	COMP VEL A
	9351	N	101-200	TU	COMP VEL B
617	MULT VEL				
	1338	E	401-500	MW	
	1349	E	101-200	FR 3	
	1543	N	101-200	TU	
	1444	E	NO INFO	BB38	
620	MULT VEL				
	1111	E	201-300	MW	
	1175	E	101-200	FR 3	
	1250	N	101-200	TU	
	1010	N	301-400	DA 2	
	1199	N	21-CM	DL 2	
	1260	N	21-CM	AL 2	
633	MULT VEL				
	1390	N	21-CM	FT	
	1370	N	21-CM	BA 4	
•637					
	880	E	401-500	L	
•663					
	1261	A	201-300	MW	
666	MULT VEL				
	28	E	401-500	L	
	38	N	21-CM	FT	
	52	N	21-CM	DD	
	45	N	21-CM	RR	
674					
	9711	E	101-200	FR11	
685	MULT VEL				
	1228	E	401-500	MW	
	1223	N	NO INFO	HJ	
•688					MAG SHARED
	1916	N	21-CM	FT	
717					
	1866	E	401-500	L	
720	MULT VEL				
	1620	E	1-100	WK 2	
	1371	E	NO INFO	DP 3	
	1410	E	101-200	WD 1	
	1374	E	101-200	DP 1	
	1620	N	21-CM	BG 4	
721					
	506	E	101-200	SN 8	
731					
	1460	E	401-500	MW	
•742	MULT VEL				
	959	E	301-400	V 8	
	1272	E	101-200	SN 8	
•744	MULT VEL				
	3302	E	101-200	SA 3	
	3600	E	201-300	AD 2	
•747					
	7200	E	201-300	AD 4	
•752					
	7588	N	101-200	TU	
•753					
	1469	E	401-500	L	
•764					
	584	N	21-CM	FT	
•766					
	2962	A	101-200	SN 8	

UGC	NGC	R.A. (1950) DEC	m_{pg}	Hubble Type	Char.	Cluster	Group
5767	N3301	10 34.2 +22 09	12.2	SO		185	
5773	N3303	10 34.4 +18 24	14.5	PEC			
5776		10 34.5 +64 32	14.4	COMP	MARK		
5786	N3310	10 35.7 +53 45	11.0	SB PEC		163	
5789	N3319	10 36.2 +41 57	12.0	SC			
5794	N3320	10 36.6 +47 39	13.1	SC			
5826	N3338	10 39.5 +14 00	12.1	SC			6346
5829		10 39.8 +34 43	15.1	IRR			
5833		10 40.3 +20 41	15.1	SO	MARK		
5837	N3329	10 40.4 +77 05	12.9	SA		172 BR IN GRP	5837
5840	N3344	10 40.7 +25 10	11.1	SC			
5842	N3346	10 41.0 +15 08	12.8	SB-C			
5846		10 41.2 +60 38	15.4	IRR			
5850	N3351	10 41.4 +11 58	11.2	SB			6346
5860	N3353	10 42.2 +56 13	12.9	PEC	MARK	189	
5873	N3359	10 43.3 +63 30	11.0	SC			
5875	N3348	10 43.4 +73 06	12.0	E		184	
5880	N3367	10 43.9 +14 02	12.0	SC			
5882	N3368	10 44.1 +12 05	10.0	SA-B SA			6346

68

UGC	V_r (km s^{-1})	Spectrum	Disp. (A/mm)	Ref.		Notes	
5767							
	1333	E	401-500	MW			
5773	MULT VEL						
	6060	E	201-300	K		COMP VEL	A
	6470	E	301-400	A	1	COMP VEL	A
	6317	E	201-300	K		COMP VEL	B
	6397	E	301-400	A	1	COMP VEL	B
5776	MULT VEL						
	1620	E	201-300	UL10			
	600	E	201-300	AD 2			
						DISCREPANT VELOCITIES?	
5786	MULT VEL						
	1039	E	401-500	MW			
	986	E	1-100	W			
	998	E	401-500	L			
	1065	N	101-200	DA 2			
	1050	E	101-200	BE 1			
	981	N	21-CM	BG 3			
	970	N	21-CM	PS			
	1021	N	NO INFO	HJ			
5789	MULT VEL						
	826	E	401-500	L			
	760	N	21-CM	GU			
	744	N	21-CM	S 3			
	742	N	21-CM	R 3			
	749	N	NO INFO	HJ			
5794							
	2331	N	21-CM	S 3			
5826							
	1330	A	401-500	L			
5829							
	633	N	21-CM	FT			
5833	MULT VEL						
	1170	E	101-200	DS 3			
	1200	E	201-300	AD 7			
5837	MULT VEL						
	1689	E	201-300	CR 2			
	1918	A	101-200	SN 8			
5840	MULT VEL						
	579	E	401-500	MW			
	589	N	21-CM	GU			
	581	N	21-CM	R 3			
	576	N	NO INFO	HJ			
5842							
	1070	E	101-200	SN 8			
5846							
	1018	N	21-CM	FT			
5850	MULT VEL						
	796	E	1-100	FR 8			
	794	E	301-400	LB			
	779	E	1-100	FR10			
5860	MULT VEL						
	1020	E	1-100	WK 2			
	862	E	101-200	DP 1			
	940	E	101-200	SN 8			
5873							
	1008	E	401-500	L			
5875							
	2855	A	401-500	MW			
5880							
	2879	E	401-500	MW		4C14.37	
5882	MULT VEL						
	927	A	201-300	MW			
	924	E	301-400	LB			
	940	N	NO INFO	SB			
	966	N	21-CM	DL 2			
	831	N	21-CM	BA 2			

UGC	NGC	R.A. (1950) DEC	m_{pg}	Hubble Type	Char.	Cluster	Group
5887	N3370	10 44.4 +17 32	12.4	SC			
5889		10 44.7 +14 20	15.0	SC-I			6346
5899	N3377	10 45.1 +14 15	10.7	E			6346
5902	N3379	10 45.2 +12 51	9.6	E			6346
5911	N3384	10 45.7 +12 54	10.0	SO			6346
5912		10 45.7 +26 51	14.5	SC	DISR	192	5884
5914	N3389	10 45.8 +12 48	12.0	SC			
5918		10 46.1 +65 50	17.0	IRR			
5931	N3395	10 47.0 +33 14	12.1	SC	DISR BR IN GRP		5931
5935	N3396	10 47.1 +33 15	12.6	IRR	DISR		5931
5943		10 47.7 -01 00	14.8	SB-C			
5947		10 47.8 +19 55	15.0	PEC			
5952	N3412	10 48.3 +13 41	10.8	SO			6346
5953		10 48.4 +44 50	13.2	PEC	MARK		
5959	N3414	10 48.5 +28 15	12.1	SO PEC		192 BR IN GRP	5959
5962	N3423	10 48.6 +06 06	12.1	SC			
5964	N3419	10 48.7 +14 14	13.4	SO			
5969	N3415	10 48.8 +43 59	13.2	SO-A E			
5977	N3408	10 49.1 +58 43	14.1	S... SC			
5978	N3407	10 49.1 +61 39	14.8	E-SO			
5981	N3433	10 49.4 +10 25	13.6	SC			6346
5982	N3430	10 49.4 +33 13	12.2	SC			5931
5984		10 49.5 +30 20	14.6	DBLE	DISR		
5986	N3432	10 49.7 +36 53	11.7	... SC			
5995	N3437	10 49.9 +23 12	12.6	SC			

UGC	V_r (km s^{-1})	Spectrum	Disp. (Å/mm)	Ref.	Notes
5887					
5889	1400	E	401-500	L	
5899	571	N	21-CM	FT	
5902	718	A	401-500	MW	
	MULT VEL				
	862	A	201-300	MW	
5911	963	A	401-500	LB	
	MULT VEL				
	781	A	201-300	MW	
5912	699	A	301-400	LB	
5914	6295	E	101-200	FR11	
	MULT VEL				
	1334	E	401-500	L	
5918	1257	E	101-200	FR 1	
					MAG NOT ZW
5931	336	N	21-CM	FT	
	MULT VEL				
	1644	E	301-400	P	
	1751	E	401-500	L	
	1622	E	1-100	FR 8	VEL CONTAM
	1622	E	1-100	DO	
5935	1631	N	21-CM	S 3	
	MULT VEL				
	1651	E	301-400	P	
	1643	E	401-500	L	
	1622	E	1-100	FR 8	VEL CONTAM
5943	1712	E	1-100	DO	
5947	4531	E	101-200	FR11	
5952	1253	N	21-CM	FT	
5953	861	A	401-500	MW	
5959	1800	E	201-300	AD 2	
	MULT VEL				
	1449	A	401-500	MW	
5962	1414	N	21-CM	BI 1	
	MULT VEL				
	1013	N	21-CM	S 3	
5964	835	A	101-200	SN 8	
	MULT VEL				
	2982	A	401-500	L	
5969	3021	N	21-CM	BI 1	
5977	3177	A	101-200	SN 8	
5978	9574	A	101-200	SN 8	
5981	5040	A	201-300	WL	
5982	2591	A	101-200	SN 8	
	MULT VEL				
	1504	E	101-200	FR 8	
	1742	E	401-500	L	
5984	1583	N	21-CM	FR 8	
5986	10808	E	301-400	KR	
	MULT VEL				
	609	E	401-500	L	
	670	N	101-200	BE 2	
5995	641	N	21-CM	R 3	
	1119	E	201-300	SN 8	

71

72

UGC	NGC	R.A. (1950) DEC	m pg	Hubble Type	Char.	Cluster	Group
5997	N3403	10 50.0 +73 57	13.3	SC		184	
5998		10 50.2 +50 34	14.5	PEC	MARK		
6001	N3442	10 50.3 +34 10	13.2	PEC	MARK		
6006	N3447	10 50.7 +17 02	14.3	... SC-I			
6007	N3447	10 50.7 +17 02	14.3	IRR			
6016		10 51.2 +54 34	17.0	IRR			
6021	N3445	10 51.5 +57 15	12.8	IRR		BR IN GRP	6021
6024	N3448	10 51.6 +54 35	12.2	... IRR			
6026	N3454	10 51.8 +17 36	14.1	... SC			
6028	N3455	10 51.9 +17 33	13.1	SB SC			
6029		10 52.0 +49 59	14.0	PEC	MARK		6013
6037	N3458	10 53.0 +57 24	13.2	SO		229	6021
6057		10 55.7 +01 54	14.9	TRPL E		197	
6060	N3470	10 55.7 +59 47	14.3	SA SC		229	
6064	N3471	10 56.0 +61 48	13.0	SA	MARK		
6069	N3478	10 56.6 +46 23	13.7	SB SC		203	
6077	N3485	10 57.4 +15 06	12.8	SB SB-C			6346
6079	N3486	10 57.6 +29 15	11.2	SC			
6082	N3489	10 57.7 +14 10	10.9	SO E			6346
6093		10 58.2 +11 00	14.8	SB-C SC		202	6094
6098	N3495	10 58.7 +03 54	13.1	SC			
6103		10 59.1 +45 30	13.4	PEC	MARK	203	
6118	N3504	11 00.4 +28 15	11.5	SA-B SB SB		206	
6119		11 00.6 +03 36	14.4	... S...			

UGC	V_r (km s^{-1})	Spectrum	Disp. (A/mm)	Ref.	Notes
5997					
	1244	E	401–500	L	
5998	MULT VEL				
	1223	E	201–300	UL10	
	1500	E	201–300	AD 2	
6001	MULT VEL				
	1620	E	101–200	DS 3	
	1800	E	201–300	AD 7	
6006	MULT VEL				MAG SHARED
	1067	E	101–200	FR 8	
	965	E	301–400	LB	
	1062	N	21–CM	FR 8	
6007					MAG SHARED
	1014	E	301–400	LB	
6016					MAG NOT ZW
	1480	N	21–CM	BG10	
6021					
	1984	E	301–400	V 5	
6024	MULT VEL				
	1404	E	301–400	V 5	
	1333	E	1–100	V 6	
	1399	N	NO INFO	HJ	
	1333	E	201–300	BG10	
	1350	N	21–CM	BG10	
	1380	N	21–CM	PS	
	1321	N	21–CM	DL 2	
6026					
	1149	E	301–400	P	
6028					
	1104	E	301–400	P	
6029	MULT VEL				
	1366	E	201–300	UL10	
	1500	E	201–300	AD 2	
6037					
	1770	A	101–200	SN 8	
6057					
	11460	A	101–200	PT	
6060					
	6639	A	101–200	SN 8	
6064	MULT VEL				
	2047	E	101–200	SA 3	
	2067	E	201–300	UL10	
	2100	E	201–300	AD 2	
6069					
	6658	A	101–200	SN 8	
6077					
	1488	E	101–200	SN 8	
6079	MULT VEL				
	1116	A	401–500	MW	
	708	N	21–CM	BA 1	
6082	MULT VEL				
	692	E	401–500	MW	
	711	E	NO INFO	BB38	
	688	N	NO INFO	HJ	
	588	N	21–CM	BI 1	
6093					
	10793	E	101–200	FR11	
6098					
	994	E	301–400	V 5	
6103	MULT VEL				
	5912	E	101–200	SA 3	
	6000	E	201–300	AD 2	
	5990	N	21–CM	BG 4	
6118	MULT VEL				
	1539	E	301–400	BB11	
	1513	E	401–500	MW	
6119					
	7454	N	101–200	TU	

UGC	NGC	R.A. (1950) DEC	m pg	Hubble Type	Char.	Cluster	Group
6120	N3506	11 00.6 +11 21	12.9	SC SB			
6126	N3510	11 00.9 +29 10	13.6	... SC		204	
6128	N3512	11 01.3 +28 18	12.9	SC		206	
6132		11 01.7 +38 28	13.1	COMP	MARK		
6134	N3509	11 01.8 +05 06	14.0	S... PEC			
6150	N3521	11 03.3 +00 15	10.1	SB			
6151		11 03.3 +20 06	17.0	SC-I			
6153	N3516	11 03.3 +72 49	12.3	SO	SEYF	212	
6175		11 04.7 +18 42	14.7	DBLE E	DISR		
6204		11 07.1 +24 32	14.5	SO S... E		207	
6207		11 07.2 +24 31	14.6	SB SC		207	
6209	N3547	11 07.3 +11 00	12.8	S... SC		209	6346
6214	N3550	11 07.9 +29 02	14.2	PEC		206	
6215	N3549	11 08.0 +53 40	12.8	SC			
6224	N3561	11 08.5 +28 58	14.7	DBLE E-SO SO-A		206	
6225	N3556	11 08.5 +55 56	10.7	SC			
6251		11 10.5 +53 52	15.7	SC-I			
6257	N3577	11 10.9 +48 33	14.7	SB			
6263	N3583	11 11.3 +48 36	11.6	SB			
6272	N3593	11 12.0 +13 05	11.8	SO			6346

UGC	V_r (km s^{-1})	Spectrum	Disp. (A/mm)	Ref.	Notes
6120					
	6459	E	101-200	SN 8	
6126	MULT VEL				
	719	E	401-500	L	
	709	N	21-CM	BG 2	
6128					
	1502	E	401-500	L	
6132					B2 1101+38
	9240	A	1-100	UL14	
6134					
	7600	E	201-300	BB28	
6150	MULT VEL				
	789	A	201-300	MW	
	815	E	101-200	BB31	
	730	N	NO INFO	SB	
	845	N	21-CM	GU	
6151					MAG NOT ZW
	1330	N	21-CM	FT	
6153	MULT VEL				
	2505	E	201-300	K	
	2614	A	201-300	MW	
	2632	E	401-500	L	
	2614	E	301-400	V 8	
	2503	N	21-CM	BA 2	
	2616	E	201-300	BX	
6175	MULT VEL				
	7982	E	101-200	ZH 1	COMP VEL A
	8211	E	101-200	ZH 1	COMP VEL B
6204	MULT VEL				
	6301	E	301-400	P	
	6173	N	101-200	TU	
6207	MULT VEL				
	5983	E	301-400	P	
	6149	N	101-200	TU	
6209					
	1513	E	101-200	SN 8	
6214	MULT VEL				
	10530	A	101-200	PT	
	10471	N	NO INFO	SN 3	
6215					
	2817	A	101-200	SN 8	
6224	MULT VEL				
	8550	E	101-200	ZH 2	COMP VEL A
	8579	E	NO INFO	SK 1	COMP VEL A
	8803	E	101-200	ZH 2	COMP VEL B
	8939	E	NO INFO	SK 1	COMP VEL B
6225	MULT VEL				
	748	E	301-400	BB 7	
	636	E	401-500	MW	
	650	E	401-500	L	
	670	N	21-CM	RR	
	700	N	21-CM	GU	
	699	N	21-CM	S 3	
	694	N	21-CM	R 3	
6251					
	928	N	21-CM	FT	
6257					
	5219	N	101-200	TU	
6263	MULT VEL				5C2.203
	2160	A	301-400	T	
	4834	N	101-200	TU	
	2082	A	101-200	SN 8	DISCREPANT VELOCITIES?
6272	MULT VEL				
	547	E	401-500	MW	
	630	E	101-200	UL 8	
	668	E	NO INFO	BB38	
	617	N	21-CM	BI 1	

UGC	NGC	R.A. (1950) DEC	m_{pg}	Hubble Type	Char.	Cluster	Group
6277	N3596	11 12.5 +15 03	11.7	SC			6346
6295	N3605	11 14.1 +18 17	12.7	E-S0			6297
6297	N3607	11 14.3 +18 19	10.2	E		BR IN GRP	6297
6299	N3608	11 14.4 +18 25	11.7	E			6297
6305	N3611	11 14.9 +04 50	12.4	SA	DISR		6368
6310	N3609	11 15.2 +26 54	14.1	SA-B		206	
6311		11 15.4 −01 49	14.3	SC		215A	
6315		11 15.4 +54 02	15.2	DBLE	MARK	214	
6318	N3614	11 15.5 +46 02	12.7	SC		216	
6319	N3610	11 15.5 +59 05	11.4	S0		229	
6321	N3612	11 15.6 +26 54	15.0	SC-I		206	
6322		11 15.6 +28 33	14.7	S0		215	
6323	N3613	11 15.6 +58 17	11.6	E-S0		229	
6328	N3623	11 16.3 +13 22	9.6	SA			6346
6330	N3619	11 16.5 +58 02	12.6	S0-A		229	
6343	N3626	11 17.5 +18 38	11.2	S0-A SA			
6345		11 17.6 +02 48	14.4	IRR		218	6368
6346	N3627	11 17.6 +13 16	8.9	SB		BR IN GRP	6346
6349	N3630	11 17.7 +03 14	12.8	S0-A		218	6368
6350	N3628	11 17.7 +13 52	11.5	S0 SB	DISR		6346
6352	N3629	11 17.9 +27 15	12.9	SC		206	
6360	N3631	11 18.2 +53 27	11.0	SC		229	
6368	N3640	11 18.5 +03 30	11.8	E		218 BR IN GRP	6368
6376	N3646	11 19.0 +20 26	11.5	SC			

UGC	V_r (km s^{-1})	Spectrum	Disp. (A/mm)	Ref.	Notes
6277	MULT VEL				
	1160	N	21-CM	BA 1	
	1176	E	101-200	SN 8	
6295					
	693	A	401-500	MW	
6297	MULT VEL				
	951	A	401-500	MW	
	871	A	401-500	L	
6299					
	1210	A	401-500	MW	
6305					
	1754	E	401-500	MW	
6310					
	5673	N	101-200	TU	
6311					
	7446	E	101-200	FR11	
6315	MULT VEL				
	10730	E	101-200	SA 1	COMP VEL A
	10959	E	101-200	SA 1	COMP VEL B
6318					
	2263	A	101-200	SN 8	
6319					
	1765	A	401-500	MW	
6321					
	5521	N	101-200	TU	
6322					
	9810	A	101-200	KZ	
6323					
	2054	A	401-500	MW	
6328	MULT VEL				
	705	E	401-500	MW	
	620	A	301-400	LB	
	814	E	301-400	BB15	
	800	N	NO INFO	SB	
6330					
	1649	E	401-500	MW	
6343	MULT VEL				
	1452	A	401-500	MW	
	1562	N	21-CM	KG 1	
6345					
	1609	N	21-CM	FT	
6346	MULT VEL				
	744	A	201-300	MW	
	638	E	301-400	LB	
	650	N	NO INFO	SB	
	720	N	21-CM	GU	
	740	N	21-CM	RO	
	778	A	101-200	SN 8	
6349	MULT VEL				
	1537	A	201-300	CR 2	
	1484	A	101-200	SN 8	
6350	MULT VEL				
	842	E	401-500	L	
	842	N	21-CM	R 3	
	855	N	21-CM	RO	
6352	MULT VEL				
	1508	N	21-CM	S 3	
	1558	E	101-200	SN 8	
6360	MULT VEL				
	1180	E	101-200	FR 8	
	1087	A	401-500	L	
	1165	N	21-CM	R 3	
6368					
	1354	A	401-500	MW	
6376	MULT VEL				
	4185	E	301-400	BB16	
	4425	E	401-500	L	
	4524	E	301-400	KR	

UGC	NGC	R.A. (1950) DEC	m pg	Hubble Type	Char.	Cluster	Group
6385	N3642	11 19.4 +59 22	11.9	SB-C SB		229	
6386	N3649	11 19.6 +20 28	14.7	S0			
6396	N3655	11 20.3 +16 52	11.6	S... SC			
6399		11 20.6 +51 12	14.9	... SC-I			
6403	N3656	11 20.8 +54 07	13.4	PEC E		229	
6405	N3659	11 21.1 +18 05	12.7	SC-I SC			
6419	N3664	11 21.8 +03 36	13.6	SB		218	6368
6420	N3666	11 21.8 +11 37	12.5	SC			6346
6426	N3665	11 22.0 +39 02	11.6	E-S0 S0			
6429		11 22.4 +64 01	14.1	SC		219	
6439	N3675	11 23.4 +43 52	10.4	SB			
6445	N3681	11 23.8 +17 08	12.2	SB-C SB			6460
6446		11 23.8 +54 01	14.5	SC		229	
6447	I 691	11 23.9 +59 26	14.2	COMP	MARK	229	
6448		11 23.9 +64 25	14.9	PEC	MARK		
6453	N3684	11 24.6 +17 18	12.1	SC			6460
6456		11 24.6 +79 16	14.7	PEC	DISR		
6458	N3683	11 24.7 +57 10	12.7	... SC		229	
6460	N3686	11 25.1 +17 30	11.6	SB-C SB		BR IN GRP	6460
6463	N3687	11 25.3 +29 47	13.0	SB SA		222	
6467	N3689	11 25.6 +25 56	12.9	SC			
6471	I 694	11 25.7 +58 50	11.8	DBLE SB	MARK	229	
6472	N3690	11 25.7 +58 50	11.8	DBLE SB	MARK	229	
6495		11 27.0 +22 24	14.9	SB SC		220	
6498	N3705	11 27.5 +09 31	11.5	SB SA-B		223	6346

78

UGC	V_r (km s^{-1})	Spectrum	Disp. (A/mm)	Ref.	Notes
6385					
	1623	E	401-500	M W	
6386					
	4689	E	301-400	K R	
6396					
	1462	E	101-200	SN 8	
6399					
	797	N	21-CM	F T	
6403					
	2803	E	201-300	K	
6405					
	1324	E	101-200	SN 8	
6419	MULT VEL				
	1406	E	301-400	V 5	
	1379	N	21-CM	F T	
6420					
	1047	E	101-200	SN 8	
6426					
	2002	A	401-500	M W	
6429					
	3724	E	101-200	FR11	
6439	MULT VEL				
	742	A	301-400	L B	
	688	A	401-500	M W	
	756	E	101-200	K T 4	
6445					
	1314	A	401-500	M W	
6446					
	2895	E	1-100	A 11	
6447	MULT VEL				
	1461	E	201-300	UL10	
	1200	E	201-300	A D 2	
	1230	E	NO INFO	WD 2	
6448	MULT VEL				
	953	E	101-200	SA 3	
	1050	E	201-300	A D 4	
6453					
	1422	A	401-500	M W	
6456					
	-110	E	1-100	C A	
6458					
	1656	E	101-200	SN 8	
6460					
	1022	E	401-500	M W	
6463					
	2377	A	101-200	SN 8	
6467	MULT VEL				4C25.35
	2640	E	201-300	W L	
	2738	E	101-200	SN 8	
6471	MULT VEL				MAG SHARED
	3111	E	301-400	V 5	
	2973	E	201-300	UL10	
	3160	E	201-300	A S	
6472	MULT VEL				MAG SHARED
	2996	E	301-400	V 5	
	2973	E	201-300	UL10	
	2973	E	101-200	SA 3	
	3000	E	NO INFO	WD 2	
	3040	E	201-300	A S	
6495					
	6547	N	101-200	FR11	
6498					
	1054	A	101-200	SN 8	

UGC	NGC	R.A. (1950) DEC			m_{pg}	Hubble Type	Char.	Cluster	Group
6503	I 701	11 28.2	+20	45	14.7	S... PEC		234	
6514		11 29.1	+71	05	15.2	MULT			
6520		11 29.6	+62	48	14.1	S...	MARK	219	
6523	N3720	11 29.8	+01	05	13.7	...E		231	
6524	N3718	11 29.8	+53	21	11.8	SO-A	DISR	229	
6527		11 29.9	+53	14	14.7	TRPL	SEYF	229	6527
6537	N3726	11 30.6	+47	18	11.2	SC			
6541		11 30.8	+49	31	13.9	IRR	MARK	226	
6542	N3725	11 30.8	+62	10	13.6	SC	MARK		
6547	N3729	11 31.1	+53	25	12.2	S... PEC		229	
6563	N3737	11 32.8	+55	13	13.9	...SO-A		229	
6565	N3738	11 33.0	+54	47	11.5	IRR		229	
6567	N3735	11 33.0	+70	48	12.4	SC			
6579	N3756	11 34.0	+54	34	12.1	SC		229	
6583		11 34.3	+20	14	13.9	PEC SB	MARK	234 BR IN GRP	6583
6589	N3768	11 34.6	+18	07	13.7	SO			
6595	N3769	11 35.0	+48	10	11.7	SB			
6597	N3746	11 35.1	+22	17	15.3	SA-B		234	6602
6605	N3773	11 35.6	+12	24	13.1	SO PEC			
6615	N3780	11 36.6	+56	32	12.2	SC		229	
6618	N3782	11 36.7	+46	47	13.1	IRR			
6621	N3786	11 37.0	+32	11	13.0	SA	DISR		
6623	N3788	11 37.1	+32	13	13.2	S...			
6630	N3799	11 37.6	+15	36	14.4	S... SA	DISR		

UGC	V_r (km s^{-1})	Spectrum	Disp. (A/mm)	Ref.	Notes
6503					
	6182	E	301-400	A 1	
6514	MULT VEL				
	15580	E	301-400	BB28	
	15690	A	1-100	SA 6	COMP VEL A
	15820	A	101-200	SA 6	COMP VEL C
	16070	A	101-200	SA 6	COMP VEL D
6520					
	3648	A	101-200	SA 3	
6523					
	5986	E	101-200	SN 8	
6524	MULT VEL				
	1050	E	401-500	MW	
	1004	N	21-CM	DL 2	
	962	N	21-CM	BG 3	
	1074	N	21-CM	R 3	
	990	N	21-CM	P S	
6527	MULT VEL				MAG SHARED
	8014	E	101-200	SA 3	
	7800	E	201-300	AD 2	
	7952	E	301-400	SA 2	
	7900	E	301-400	BB12	COMP VEL A
	6700	E	301-400	BB12	COMP VEL B
					DISCREPANT VELOCITIES?
6537	MULT VEL				
	948	E	401-500	MW	
	762	N	21-CM	R 3	
6541	MULT VEL				
	218	E	101-200	SA 3	
	218	E	201-300	UL10	
	200	N	21-CM	BG 4	
6542	MULT VEL				
	3180	A	201-300	UL10	
	3254	A	201-300	SN 8	
6547					
	1005	E	101-200	SN 8	
6563					
	5580	A	101-200	PT	
6565					
	146	E	101-200	SN 8	
6567					
	2597	E	101-200	SN 8	
6579					
	1071	E	201-300	SN 8	
6583	MULT VEL				
	6191	E	101-200	SA 3	
	6190	E	NO INFO	SA 2	
	6300	E	201-300	AD 4	
	6201	N	101-200	K I	
6589					
	3301	N	101-200	K I	
6595					
	737	E	301-400	P	
6597					
	9074	N	NO INFO	BN	
6605					
	984	E	101-200	SN 8	
6615					
	2361	E	201-300	SN 8	
6618					
	728	N	21-CM	FT	
6621					
	2741	E	301-400	P	
6623					
	2326	E	301-400	P	
6630					
	3545	E	301-400	P	

UGC	NGC	R.A. (1950) DEC	m$_{pg}$	Hubble Type	Char.	Cluster	Group
6634	N3800	11 37.7 +15 37	13.1	S... SB			
6635	N3801	11 37.7 +18 00	13.3	SO E		BR IN GRP	6635
6637		11 37.8 +28 40	14.5	...		230	
6642	N3805	11 38.2 +20 38	13.8	E-SO E		234	
6643	N3808	11 38.2 +22 43	14.1	DBLE S...	DISR	234	
6644	N3810	11 38.3 +11 45	11.4	SC.			6346
6650	N3811	11 38.6 +47 58	13.0	SC	MARK		
6651	N3813	11 38.7 +36 49	12.6	S...		235	
6656	N3816	11 39.2 +20 23	13.6	SC SO E	DISR	234	
6663	N3821	11 39.6 +20 36	13.8	SA SO		234	
6667		11 39.7 +51 53	14.8	SC			
6673	N3827	11 40.0 +19 07	13.6	... S...		234	
6682		11 40.5 +59 23	16.5	SC-I		229	
6683		11 40.6 +20 01	15.2	SO-A SO		234	
6688	I2951	11 40.8 +20 01	15.0	SA		234	
6693	N3832	11 40.9 +23 00	14.0	SC S...		234	
6697		11 41.2 +20 15	14.3	IRR		234	
6701	N3837	11 41.3 +20 10	14.2	S... E.		234	
6702	N3840	11 41.3 +20 21	14.7	SA		234	
6704	N3842	11 41.4 +20 13	13.3	S..:. E.		234 BR IN CLUS	
6705	N3844	11 41.4 +20 18	14.9	SO-A		234	
6718	N3860	11 42.2 +20 05	14.5	SO SA-B		234	
6719		11 42.2 +20 24	14.6	S...		234	
6721	N3859	11 42.3 +19 44	14.9	PEC S...		234	

UGC	V_r (km s^{-1})	Spectrum	Disp. (Å/mm)	Ref.	Notes
6634					
	3528	E	301-400	P	
6635	MULT VEL				4C17.52
	3150	E	201-300	WL	
	3254	N	101-200	KI	
6637					
	2970	E	101-200	DP 1	
6642	MULT VEL				
	6697	A	101-200	DM	
	6472	N	101-200	KI	
6643					
	7050	E	101-200	A 4	
6644	MULT VEL				
	1005	E	401-500	L	
	972	A	401-500	MW	
	900	E	201-300	BS	
6650	MULT VEL				
	3042	A	101-200	SA 3	
	3000	E	201-300	AD 4	
6651	MULT VEL				
	1386	E	201-300	CR 2	
	1484	A	101-200	SN 8	
6656	MULT VEL				
	5742	A	201-300	DM	
	5548	N	101-200	KI	
6663	MULT VEL				
	5807	A	201-300	DM	
	5535	N	101-200	KI	
6667					
	975	N	21-CM	FT	
6673	MULT VEL				
	3143	E	201-300	DM	
	3268	N	101-200	KI	
6682					MAG NOT ZW
	1318	N	21-CM	FT	
6683					
	7526	A	201-300	DM	
6688					
	7646	A	101-200	GD	
6693					
	6906	E	201-300	DM	
6697	MULT VEL				
	6589	E	101-200	GD	
	6698	E	201-300	DM	
6701	MULT VEL				
	6353	A	101-200	GD	
	6226	A	201-300	DM	
6702	MULT VEL				
	7417	E	101-200	GD	
	7352	E	201-300	DM	
6704	MULT VEL				
	6180	A	101-200	PT	
	6164	E	101-200	GD	
	6111	N	101-200	KI	
6705	MULT VEL				
	6824	A	101-200	GD	
	6868	A	101-200	DM	
6718					
	5461	E	101-200	GD	
6719	MULT VEL				
	6724	A	101-200	DM	
	6590	E	101-200	GD	
6721					
	5508	E	201-300	DM	

UGC	NGC	R.A. (1950) DEC	m_{pg}	Hubble Type	Char.	Cluster	Group
6723	N3862	11 42.5 +19 53	14.0	E		234	
6724	N3861	11 42.5 +20 15	14.0	SB S...		234	
6732		11 42.8 +59 15	13.5	...		229	
6735	N3873	11 43.1 +20 03	14.2	E		234	
6738	N3872	11 43.2 +14 03	12.9	E			
6742	N3870	11 43.3 +50 30	13.2	SO	MARK		
6745	N3877	11 43.5 +47 46	11.8	SC			
6746	N3884	11 43.6 +20 41	14.0	SA SO		234	
6754	N3883	11 44.2 +20 58	14.2	SB S...		234	
6760	N3886	11 44.5 +20 07	14.3	E-SO E		234	
6765	N3888	11 44.9 +56 14	12.6	SC	MARK	229	6787
6778	N3893	11 46.0 +48 59	10.6	SB-C SC	DISR		
6782		11 46.4 +24 07	17.0	IRR		234	
6784	N3897	11 46.4 +35 18	14.2	SB-C SC		235	
6786	N3900	11 46.5 +27 19	12.5	SA			
6787	N3898	11 46.5 +56 20	11.7	SA		229 BR IN GRP	6787
6801	N3912	11 47.4 +26 45	13.2	SC-I SA			
6813	N3913	11 48.0 +55 37	14.2	SC		229	
6815	N3917	11 48.1 +52 07	12.5	SC		229	
6816		11 48.1 +56 44	15.1	IRR		229	
6817		11 48.2 +39 09	15.1	IRR			
6818		11 48.2 +46 05	14.6	...	DISR		
6823	N3921	11 48.5 +55 21	13.4	PEC	MARK	229	
6833	N3930	11 49.1 +38 16	13.5	SC			
6834	N3928	11 49.1 +48 57	13.1	...	MARK		

84

UGC	V_r (km s^{-1})	Spectrum	Disp. (A/mm)	Ref.	Notes
6723	MULT VEL				3C264
	6468	A	101-200	GD	
	6592	A	201-300	DM	
	6240	A	301-400	ST	
	6706	N	101-200	KI	
6724	MULT VEL				
	5040	A	101-200	DM	
	5015	E	101-200	GD	
6732	3090	A	301-400	CA	
6735	5552	A	101-200	GD	
6738	3109	A	401-500	MW	
6742	MULT VEL				
	631	E	101-200	SA 3	
	741	E	101-200	UL10	
	600	E	201-300	AD 2	
	755	N	21-CM	BG 4	
6745	838	E	201-300	SN 8	
6746	MULT VEL				
	7002	E	101-200	DM	
	6822	N	101-200	KI	
6754	7406	A	201-300	DM	
6760	5431	A	101-200	DM	
6765	MULT VEL				
	2419	A	101-200	SA 3	
	2400	E	201-300	AD 4	
	2019	A	201-300	SN 8	
6778	MULT VEL				
	932	E	101-200	FR 8	
	1042	E	401-500	MW	
	868	A	401-500	L	
	980	N	21-CM	FR 8	
6782	528	N	21-CM	FT	MAG NOT ZW
6784	6484	N	101-200	FR11	
6786	1702	A	401-500	MW	
6787	1038	E	401-500	MW	
6801	1698	E	101-200	SN 8	
6813	831	N	NO INFO	SN 1	
6815	966	N	21-CM	FT	
6816	896	N	21-CM	FT	
6817	248	N	21-CM	FT	
6818	806	N	21-CM	FT	
6823	MULT VEL				
	5995	E	101-200	DS 3	
	6000	E	201-300	AD 7	
	5930	E	301-400	BB28	
6833	915	N	NO INFO	HJ	
6834	MULT VEL				
	965	E	101-200	SA 3	
	900	E	201-300	AD 4	

UGC	NGC	R.A. (1950) DEC	m_pg	Hubble Type	Char.	Cluster	Group
6840		11 49.5 +52 23	15.6	SC-I		229	
6856	N3938	11 50.2 +44 23	11.0	SC		248	
6857	N3941	11 50.3 +37 15	11.3	SO			
6860	N3945	11 50.5 +60 56	11.6	SO		229	
6863	N3947	11 50.8 +21 02	14.2	SB		234	
6869	N3949	11 51.1 +48 08	10.9	S•••• SC			
6870	N3953	11 51.2 +52 36	10.8	SB		229	
6884	N3963	11 52.4 +58 46	12.2	SB SC		229	
6886		11 52.6 +06 27	14.5	SB SC			
6899	N3971	11 53.0 +30 16	13.9	SO		244	
6906	N3976	11 53.3 +07 02	12.8	SB SC			
6909	N3977	11 53.5 +55 40	14.7	SA		229	
6910	N3978	11 53.5 +60 47	13.2	SB SC		229	
6917		11 53.9 +50 43	14.0	•••• SC-I			
6918	N3982	11 53.9 +55 23	11.6	S•••		229	
6921	N3985	11 54.1 +48 36	13.0	SB-C S•••			
6927		11 54.5 +30 40	14.5	SO		244	
6933	N3991	11 54.9 +32 36	13.8	PEC IRR		244	6944
6935	N3993	11 55.0 +25 31	14.8	S••• SO		242	6952
6936	N3994	11 55.0 +32 33	13.7	S••• SC		244	6944
6937	N3992	11 55.0 +53 39	10.7	SB		229	
6938	N3990	11 55.0 +55 43	13.6	SO		229	
6942	N3997	11 55.2 +25 33	14.3	SB SB-C	DISR	242	6952
6944	N3995	11 55.2 +32 34	12.9	S••• SC SC SC	DISR BR IN GRP	244	6944
6946	N3998	11 55.3 +55 43	11.2	SO		229	
6950	N4004	11 55.4 +28 09	14.0	PEC	MARK	243	

UGC	V_r (km s^{-1})	Spectrum	Disp. (A/mm)	Ref.	Notes
6840					
	1019	N	21-CM	FT	
6856	MULT VEL				
	742	E	101-200	FR 8	
	874	E	401-500	L	
	812	N	21-CM	R 3	
6857	MULT VEL				
	972	A	401-500	MW	
	927	A	401-500	L	
	917	N	21-CM	BI 1	
6860					
	1220	A	401-500	MW	
6863					
	6288	E	201-300	DM	
6869					
	681	A	401-500	MW	
6870	MULT VEL				
	938	A	401-500	MW	
	1008	A	401-500	L	
6884					
	3204	E	201-300	SN 8	
6886					
	6973	E	101-200	FR11	
6899					
	6880	N	NO INFO	CR 4	
6906					
	2491	A	101-200	SN 8	
6909					
	5710	A	NO INFO	KZ	
6910					
	9960	E	201-300	SN 8	
6917					
	905	N	21-CM	FT	
6918					
	945	E	201-300	SN 8	
6921					
	928	E	101-200	SN 8	
6927					
	3209	N	NO INFO	CR 4	
6933	MULT VEL				
	3302	E	301-400	P	
	3305	E	401-500	LB	
6935					
	4824	E	301-400	P	
6936					
	3118	E	301-400	P	
6937					
	1059	A	401-500	MW	
6938					
	720	A	401-500	L	
6942					
	4762	E	301-400	P	
6944	MULT VEL				
	3329	E	301-400	P	COMP VEL A
	3386	E	301-400	P	COMP VEL B
	3347	E	401-500	L	
6946	MULT VEL				
	1173	E	1-100	DC	
	1109	E	401-500	MW	
	1059	E	401-500	L	
	1118	E	301-400	UL 9	
	1185	E	NO INFO	BB38	
6950	MULT VEL				
	3440	E	101-200	DS 3	
	3300	E	201-300	AD 7	
	3447	N	NO INFO	CR 4	

UGC	NGC	R.A. (1950) DEC	m_{pg}	Hubble Type	Char.	Cluster	Group
6953	N4008	11 55.6 +28 28	13.1	E-S0		243	
6954	N4016	11 55.8 +27 48	14.6	S0 SC-I		243	
6955		11 55.8 +38 20	15.2	IRR			
6956		11 55.8 +51 13	17.0	SC-I		229	
6963	N4013	11 56.0 +44 13	12.4	SB		248	
6964	N4010	11 56.0 +47 32	13.1	SB-C SC-I			
6967	N4017	11 56.1 +27 43	13.5	SB-C	DISR	243	
6971	N4020	11 56.3 +30 41	13.2	...			
6973	I 750	11 56.3 +42 59	12.7	S...		248	
6983		11 56.6 +52 59	14.5	SC		229	
6985	N4026	11 56.8 +51 15	11.5	S0		229	
6993	N4030	11 57.8 −00 48	12.4	SB SC		231	
7002	N4037	11 58.8 +13 40	13.8	S...			
7005	N4036	11 58.9 +62 10	11.5	SB-C S0		229	7005 BR IN GRP
7014	N4041	11 59.7 +62 25	11.6	SC		229	
7020		12 00.0 +41 20	14.4	SB-C SB SC			
7020A		12 00.1 +64 39	14.3	S0	MARK	255	
7021	N4045	12 00.2 +02 16	13.5	SA SB-C		249	
7025	N4047	12 00.3 +48 55	12.8	S... SA-B			
7030	N4051	12 00.6 +44 48	11.5	SB-C SB SC	SEYF		
7032		12 00.9 +16 46	14.0	... S0-A			7032 BR IN GRP
7044	N4061	12 01.5 +20 30	14.4	E COMP		251	
7045	N4062	12 01.5 +32 10	11.9	SC SA		263	
7050	N4065	12 01.6 +20 30	14.0	E COMP		251	BR IN CLUS
7053		12 01.7 −01 15	17.0	IRR			
7054	N4064	12 01.7 +18 43	12.5	SA			

UGC	V_r (km s^{-1})	Spectrum	Disp. (A/mm)	Ref.	Notes
6953	MULT VEL				
	3603	N	NO INFO	CR 4	
	3520	A	201-300	SN 8	
6954					
	3494	N	NO INFO	CR 4	
6955					
	917	N	21-CM	FT	
6956					MAG NOT ZW
	912	N	21-CM	FT	
6963	MULT VEL				
	842	N	21-CM	FT	
	631	A	101-200	SN 8	
6964					
	906	N	21-CM	FT	
6967					
	3351	N	NO INFO	CR 4	
6971					
	815	N	NO INFO	CR 4	
6973					
	683	E	101-200	SN 8	
6983					
	1053	N	21-CM	FT	
6985					
	878	A	401-500	MW	
6993					
	1509	A	401-500	L	
7002					
	896	E	101-200	SN 8	
7005					
	1382	E	401-500	MW	
7014					
	1186	E	301-400	V 5	
7020					
	6137	E	101-200	FR11	
7020A	MULT VEL				
	1426	E	201-300	UL10	
	1386	E	101-200	SA 3	
	1500	E	NO INFO	WD 2	
7021	MULT VEL				
	1840	A	301-400	V 8	
	1971	E	101-200	SN 8	
	1842	N	101-200	EA	
7025					
	3325	A	101-200	SN 8	
7030	MULT VEL				
	627	E	201-300	MW	
	658	E	301-400	V 5	
	677	N	NO INFO	DA 1	
	657	N	NO INFO	HJ	
	646	N	301-400	DA 2	
	766	N	21-CM	DL 2	
	690	N	21-CM	AL 2	
7032	MULT VEL				
	3910	N	101-200	TU	
	4200	E	201-300	AD10	
7044					
	1604	N	101-200	TU	
7045	MULT VEL				
	748	A	201-300	CR 3	
	640	A	101-200	SN 8	
7050					
	1181	N	101-200	TU	
7053					MAG NOT ZW
	1463	N	21-CM	FT	
7054					
	1033	E	401-500	L	

UGC	NGC	R.A. (1950) DEC	m_{pg}	Hubble Type	Char.	Cluster	Group
7060	N4073	12 01.9 +02 11	13.8	E		249	
7062	N4081	12 02.0 +64 43	13.6	S...		255	
7063	N4077	12 02.1 +02 04	14.5	E-S0		249	
7064		12 02.1 +31 27	14.0	S... E		263 BR IN GRP	7064
7068	N4080	12 02.4 +27 16	14.0	IRR			
7073		12 02.8 +18 09	14.1	SC			
7074		12 02.8 +18 12	14.6	S...			
7075	N4085	12 02.8 +50 39	12.8	SC		229	
7081	N4088	12 03.0 +50 50	11.2	SC		229	
7085		12 03.2 +09 16	14.8	S...	DISR		
7085A		12 03.2 +31 20	15.1	DBLE		263	
7089		12 03.4 +43 25	14.8	SC-I		248	
7090	N4096	12 03.4 +47 45	11.6	SC			
7095	N4100	12 03.6 +49 52	11.7	SB-C SC			
7096	N4102	12 03.8 +53 00	11.8	SB		229	
7099	N4104	12 04.1 +28 27	13.7	SA S0 E		263	
7101	N4108	12 04.2 +67 26	13.0	SC		255 BR IN GRP	7101
7103	N4111	12 04.5 +43 20	11.4	S0		248	
7111	N4116	12 05.1 +02 58	13.0	... SC		500	
7116	N4123	12 05.6 +03 10	13.1	SC SB-C		500	
7117	N4124	12 05.6 +10 39	12.7	S0		500	
7118	N4125	12 05.6 +65 27	10.9	E		255	
7120	N4128	12 06.0 +69 03	12.7	S0			
7126	N4131	12 06.2 +29 35	14.1	S... E		263	7130
7130	N4134	12 06.6 +29 27	13.8	SB		263 BR IN GRP	7130

UGC	V_r (km s^{-1})	Spectrum	Disp. (A/mm)	Ref.	Notes
7060	MULT VEL				
	5931	A	201-300	SN 8	
	5756	N	101-200	EA	
7062					
	1312	N	101-200	KI	
7063					
	6921	N	101-200	TU	
7064					
	7469	E	201-300	CR 3	
7068					
	749	N	NO INFO	CR 4	
7073					
	4492	N	101-200	TU	
7074					
	4699	N	101-200	TU	
7075	MULT VEL				
	751	N	21-CM	FT	
	684	E	101-200	SN 8	
7081	MULT VEL				
	722	E	301-400	LB	
	739	E	401-500	L	
	648	A	101-200	SN 8	
7085					
	6166	N	101-200	TU	
7085A	MULT VEL				
	6894	E	101-200	ZH 1	
	7010	A	101-200	ZH 1	
7089					
	772	N	21-CM	FT	
7090	MULT VEL				
	524	E	101-200	BC	
	474	N	21-CM	BG 3	
	500	N	21-CM	BA 1	
	449	A	101-200	SN 8	
7095	MULT VEL				
	1063	N	21-CM	FT	
	1108	E	101-200	SN 8	
7096	MULT VEL				
	878	E	401-500	L	
	908	E	401-500	MW	
7099					
	8493	A	201-300	CR 3	
7101					
	2485	N	101-200	KI	
7103	MULT VEL				
	784	E	201-300	MW	
	870	E	401-500	L	
	800	N	NO INFO	SB	
7111					
	1304	E	401-500	L	
7116	MULT VEL				
	1253	E	101-200	SN 8	
	1283	N	101-200	EA	
7117	MULT VEL				
	1622	A	101-200	SN 8	
	1582	N	101-200	EA	
7118	MULT VEL				
	1305	A	401-500	MW	
	1485	A	401-500	L	
	1380	N	101-200	TU	
	1356	E	NO INFO	BB38	
7120					
	2395	A	401-500	L	
7126					
	3710	A	201-300	CR 3	
7130					
	3860	N	NO INFO	CR 4	

UGC	NGC	R.A. (1950) DEC	m_{pg}	Hubble Type	Char.	Cluster	Group
7132		12 06.6 +31 51	14.4	E		263	
7134	N4136	12 06.8 +30 12	12.1	SC		263	
7139	N4138	12 07.0 +43 57	12.1	SO SA			
7142	N4143	12 07.1 +42 48	12.0	SO		248	
7151	N4144	12 07.4 +46 43	12.3	SC			
7154	N4145	12 07.5 +40 10	12.2	SC			
7158	N4148	12 07.6 +36 09	14.6	SO		258	
7163	N4146	12 07.8 +26 42	13.8	SA			
7165	N4150	12 08.0 +30 40	12.6	SO		263	
7166	N4151	12 08.0 +39 41	11.2	SA-B SA	SEYF		
7169	N4152	12 08.1 +16 18	12.5	SC		500	
7168		12 08.0 +70 40	15.4	PEC	MARK		
7173	N4156	12 08.3 +39 45	14.3	SB SC S... SA-B			
7175		12 08.4 +40 02	15.3	IRR			
7176		12 08.4 +50 36	17.0	IRR		229	
7178		12 08.5 +02 18	17.0	IRR			
7179		12 08.5 +64 12	14.0	S...		255	
7182	N4158	12 08.6 +20 27	13.1	S... E			
7183	N4157	12 08.6 +50 47	11.9	SB		229	
7193	N4162	12 09.3 +24 24	12.6	SB-C SC		260	
7201	N4165	12 09.7 +13 31	14.8	SB		500	

UGC	V_r (km s^{-1})	Spectrum	Disp. (A/mm)	Ref.	Notes
7132					
	6758	E	201-300	CR 3	
7134					
	445	E	401-500	MW	
7139					
	1039	A	401-500	MW	
7142					
	784	A	401-500	MW	
7151	MULT VEL				
	536	N	21-CM	L S	
	263	N	21-CM	S 3	
7154	MULT VEL				
	1001	E	201-300	CR 2	
	1075	N	301-400	V 8	
	1026	N	21-CM	BL	
	799	A	101-200	SN 8	
7158					
	4750	N	NO INFO	RT	
7163					
	6488	N	NO INFO	CR 4	
7165					
	244	E	401-500	MW	
7166	MULT VEL				
	934	E	401-500	L	
	960	E	NO INFO	MW	
	947	E	1-100	OK	
	990	E	1-100	WD 1	
	970	E	1-100	UL13	
	875	N	101-200	TU	
	980	N	NO INFO	SB	
	952	N	NO INFO	DA 1	
	907	N	301-400	DA 2	
	980	E	1-100	AN 1	
	953	E	1-100	AN 3	
	980	E	101-200	FR 2	
	996	N	21-CM	D	
	1000	N	21-CM	BL	
	1005	N	21-CM	AL 2	
7169					
	2120	E	101-200	SN 8	
7168					
	2155	E	101-200	SA 3	
7173	MULT VEL				
	6765	E	101-200	FR11	
	725	N	101-200	TU	
	6848	E	101-200	SN 8	DISCREPANT VELOCITIES?
7175					MAG SHARED
	1886	N	101-200	BH	
7176					MAG NOT ZW
	859	N	21-CM	FT	
7178					MAG NOT ZW
	1339	N	21-CM	FT	
7179					
	2641	N	101-200	KI	
7182	MULT VEL				
	2506	E	101-200	SN 8	
	2200	N	101-200	EA	
7183	MULT VEL				
	805	N	21-CM	BA 1	
	878	A	201-300	BC	
	920	A	101-200	KZ	
	782	N	21-CM	BG 3	
	651	A	101-200	SN 8	
7193					
	2546	A	401-500	L	
7201					MAG SHARED
	1505	N	101-200	EA	

UGC	NGC	R.A. (1950) DEC	m_{pg}	Hubble Type	Char.	Cluster	Group
7202	N4169	12 09.7 +29 27	12.9	SO E		263	7202
7203	N4168	12 09.8 +13 29	12.7	E		BR IN GRP 500	
7204	N4173	12 09.8 +29 29	13.7	...		263	7202
7206	N4174	12 09.9 +29 25	14.3	... SO		263	7202
7209	I 769	12 10.0 +12 24	14.1	SB-C		500	
7211	N4175	12 10.0 +29 26	14.2	S... SA		263	7202
7214	N4179	12 10.3 +01 35	12.8	SO E		500	
7215	N4178	12 10.3 +11 08	12.9	SC		500	
7221		12 10.7 +29 07	15.0	S...		263	
7222	N4183	12 10.7 +43 58	13.5	SC			
7223	N4186	12 10.8 +15 03	16.0	...		500	
7225	N4185	12 10.8 +28 47	13.5	SB		263	
7231	N4192	12 11.2 +15 10	11.0	SB		500	
7235	N4189	12 11.3 +13 42	12.7	SC		500	
7241	N4194	12 11.6 +54 48	13.0	PEC SO	MARK		
7245	N4196	12 11.9 +28 42	13.7	SO E		263	
7249		12 12.1 +13 05	15.3	IRR		500	
7256	N4203	12 12.5 +33 28	11.8	SO			
7257		12 12.5 +36 14	14.2	SC-I			
7258	N4205	12 12.5 +64 04	13.8	S...		255	
7260	N4206	12 12.7 +13 18	13.8	SC		500	
7261	N4204	12 12.7 +20 56	14.3	...			
7264	N4210	12 12.8 +66 15	13.4	SB		255	
7275	N4212	12 13.1 +14 11	11.9	SC		500	

94

UGC	V_r (km s^{-1})	Spectrum	Disp. (A/mm)	Ref.	Notes
7202					
7203	3840 MULT VEL	A	201-300	CR 3	
	2398	A	201-300	SN 8	
	2248	N	101-200	EA	
7204					
7206	1106	E	NO INFO	CR 4	
7209	4161	A	201-300	CR 3	
7211	2213	N	21-CM	FT	
7214	4061	E	201-300	CR 3	
7215	1279 MULT VEL	A	401-500	MW	
	233	E	401-500	L	
	329	N	21-CM	DL 1	
	428	N	101-200	EA	
7221	3862	N	NO INFO	CR 4	
7222	937	N	21-CM	FT	
7223	2090	N	101-200	EA	MAG NOT ZW
7225	4074	N	NO INFO	CR 4	
7231	MULT VEL				
	-124	E	401-500	MW	
	-143	N	21-CM	DL 1	
	-121	N	101-200	EA	
7235	MULT VEL				
	2047	N	NO INFO	SA 4	
	2184	A	101-200	SN 8	
7241	MULT VEL				
	2585	E	401-500	L	
	2400	E	201-300	AD 3	
	2522	E	301-400	A 1	
	2495	E	101-200	UL 3	
	2545	N	21-CM	BG 4	
	2515	N	21-CM	PS	
	2530	E	201-300	AS	
7245	3982	A	201-300	CR 3	
7249	654	N	21-CM	FT	
7256	MULT VEL				
	1001	A	401-500	MW	
	1093	N	21-CM	KG 1	
	1074	N	21-CM	BI 1	
7257	3980	N	NO INFO	RT	
7258	1435	N	101-200	KI	
7260	MULT VEL				
	1368	A	301-400	V 8	
	373	N	101-200	EA	DISCREPANT VELOCITIES?
7261					
7264	741	N	NO INFO	GT	
7275	2501	N	101-200	KI	
	2125	A	401-500	L	

UGC	NGC	R.A. (1950) DEC	m pg	Hubble Type	Char.	Cluster	Group
7277	N4211	12 13.1 +28 27	14.4	DBLE SO SO SB SB		263	
7278	N4214	12 13.1 +36 37	10.3	IRR			
7281	N4215	12 13.3 +06 40	13.0	SO-A SO		500	
7282	N4217	12 13.3 +47 21	12.4	SB			
7283	N4218	12 13.3 +48 23	13.2	•••			
7284	N4216	12 13.4 +13 26	11.2	SB		500	
7288	N4221	12 13.6 +66 30	13.6	SO		255	
7290	N4220	12 13.7 +48 09	12.4	SA			
7291	N4222	12 13.9 +13 35	14.6	SC		500	
7292	N4224	12 14.0 +07 44	13.3	SA SO-A		500	
7296	N4227	12 14.0 +33 48	13.8	SO-A			
7299	N4229	12 14.1 +33 50	14.3	S•••			
7300		12 14.2 +29 00	17.0	IRR		263	
7306	N4236	12 14.4 +69 45	10.7	••• SC			
7307		12 14.5 +10 17	16.5	IRR		500	
7309	N4234	12 14.6 +03 58	13.4	IRR		500	
7310	N4235	12 14.6 +07 28	13.2	SA SB	SEYF	500	
7311	N4233	12 14.6 +07 54	13.2	SO		500	
7315	N4237	12 14.7 +15 36	12.3	SB SC		500	
7322	N4244	12 15.0 +38 05	10.8	SC			
7323	N4242	12 15.0 +45 53	11.9	••• SC-I			
7328	N4245	12 15.1 +29 53	12.4	SO-A SA		263	

UGC	V_r (km s^{-1})	Spectrum	Disp. (A/mm)	Ref.	Notes
7277	MULT VEL				
	6695	A	201-300	CR 3	COMP VEL A
	6571	E	101-200	ZH 1	COMP VEL A
	6838	A	201-300	CR 3	COMP VEL B
	6618	E	101-200	ZH 1	COMP VEL B
7278	MULT VEL				
	295	E	401-500	MW	
	318	E	401-500	L	
	300	N	NO INFO	SB	
	388	N	21-CM	RR	
	288	N	21-CM	SW	
7281					
	2063	A	101-200	SN 8	
7282	MULT VEL				
	962	N	21-CM	LS	
	985	N	21-CM	BA 1	
	962	N	21-CM	BG 3	
7283					
	1388	E	101-200	DP 1	
7284	MULT VEL				
	32	A	201-300	MW	
	59	A	401-500	L	
	-103	N	21-CM	DL 1	
	-54	N	101-200	EA	
7288					
	2816	N	101-200	KI	
7290					
	979	A	401-500	MW	
7291					
	10	N	101-200	EA	
7292	MULT VEL				
	2621	A	101-200	SN 8	
	2458	N	101-200	EA	
7296	MULT VEL				
	6516	N	101-200	TU	
	6393	N	NO INFO	CR 4	
	4820	N	NO INFO	R T	DISCREPANT VELOCITIES?
7299	MULT VEL				
	6663	N	101-200	TU	
	6765	N	NO INFO	CR 4	
7300					
	1215	N	21-CM	FT	MAG NOT ZW
7306	MULT VEL				
	27	E	401-500	L	
	5	N	21-CM	RR	
7307					
	1180	N	21-CM	FT	MAG NOT ZW
7309					
	2143	E	301-400	LB	
7310	MULT VEL				
	2408	E	101-200	AE	
	2566	E	101-200	SN 8	
7311					
	2194	A	201-300	SN 8	
7315	MULT VEL				
	915	A	101-200	SN 8	
	891	N	101-200	EA	
7322	MULT VEL				
	265	E	401-500	L	
	245	N	21-CM	RR	
	229	N	21-CM	SW	
	242	N	21-CM	HU 2	
	241	N	NO INFO	HJ	
7323					
	661	E	301-400	LB	
7328					
	890	A	401-500	MW	

UGC	NGC	R.A. (1950) DEC			m pg	Hubble Type	Char.	Cluster	Group
7338	N4251	12 15.6	+28	28	11.5	SO		263	
7340		12 15.7	+44	27	14.2	DBLE	MARK		
7344	N4253	12 15.9	+30	05	13.7	SA		263	
7345	N4254	12 16.2	+14	42	10.2	SC		500	
7351	N4256	12 16.4	+66	10	12.7	SB		255	
7353	N4258	12 16.5	+47	35	9.6	SB			
7354		12 16.6	+04	08	14.5	...	MARK	500	
7359	N4259	12 16.8	+05	39	14.5	SO		500	
7360	N4261	12 16.8	+06	06	12.0	E		500	
7361	N4260	12 16.8	+06	22	13.1	SA		500	
7363	I 777	12 16.9	+28	35	14.5	S...		263	
7364	N4264	12 17.0	+06	07	13.9	SO		500	
7365	N4262	12 17.0	+15	09	12.3	SO		500	
7371	N4268	12 17.2	+05	34	13.9	SO-A		500	
7372	N4269	12 17.2	+06	18	13.9	SO-A		500	
7373	N4267	12 17.2	+13	05	12.4	SO		500	
7376	N4270	12 17.3	+05	45	13.3	SO E		500	
7377	N4274	12 17.3	+29	54	11.1	SA		263	
7378	N4272	12 17.3	+30	27	14.2	E		263	
7380	N4273	12 17.4	+05	37	12.3	SC	DISR	500	
7382	N4275	12 17.4	+27	54	13.4	SB		263	
7384	I3165	12 17.5	+28	15	14.9	SB		263	
7386	N4278	12 17.6	+29	33	11.2	E		263	

UGC	V_r (km s^{-1})	Spectrum	Disp. (Å/mm)	Ref.	Notes
7338					
	1014	A	401-500	MW	
7340					
	7500	E	201-300	AD 3	
7344					
	3876	N	NO INFO	CR 4	
7345	MULT VEL				
	2485	A	401-500	MW	
	2451	A	401-500	L	
	2387	N	21-CM	DL 1	
7351	MULT VEL				
	2583	A	401-500	L	
	2559	N	101-200	K I	
7353	MULT VEL				
	510	E	301-400	BB27	
	420	E	201-300	MW	
	472	E	301-400	V 5	
	508	E	1-100	W	
	500	N	NO INFO	SB	
	457	N	101-200	DA 2	
	395	N	101-200	KT 3	
	450	N	21-CM	RR	
7354	MULT VEL				
	1564	E	101-200	DP 3	
	1716	E	301-400	LB	
	1260	E	101-200	WK 2	
7359					
	2426	N	101-200	EA	
7360	MULT VEL				3C270
	2202	A	401-500	MW	
	2183	N	101-200	EA	
7361	MULT VEL				
	1935	A	301-400	LB	
	1906	N	101-200	EA	
7363					
	2620	N	NO INFO	CR 4	
7364					
	2632	N	101-200	EA	
7365	MULT VEL				
	1351	A	301-400	V 5	
	1354	N	21-CM	BI 1	
7371					
	2317	N	101-200	EA	
7372					
	2534	N	101-200	EA	
7373					
	1260	A	401-500	MW	
7376	MULT VEL				
	2347	A	401-500	MW	
	2349	N	101-200	EA	
7377	MULT VEL				
	767	A	401-500	MW	
	718	N	21-CM	DL 2	
7378					
	8462	A	201-300	CR 3	
7380	MULT VEL				
	2302	E	401-500	MW	
	2421	N	101-200	EA	
7382	MULT VEL				
	2307	E	201-300	CR 3	
	2400	E	201-300	AD10	
7384					
	8443	N	NO INFO	CR 4	
7386	MULT VEL				B2 1217+29
	590	E	101-200	DC	
	624	E	401-500	MW	
	703	E	301-400	P	
	680	N	21-CM	BG 7	
	620	N	21-CM	BI 1	

99

UGC	NGC	R.A. (1950) DEC				m_{pg}	Hubble Type	Char.	Cluster	Group
7389	N4281	12	17.8	+05	40	12.5	SO		500	
7390	N4283	12	17.8	+29	35	13.1	E		263	
7395		12	17.9	+31	27	14.5	...		263	
7397	N4291	12	18.0	+75	40	12.3	E		269	
7399	N4288	12	18.2	+46	35	13.6	SC		265	
7402	N4290	12	18.4	+58	22	12.8	SB			
7405	N4293	12	18.7	+18	40	11.6	SO-A SA		500	
7407	N4294	12	18.8	+11	47	12.6	SC		500	
7408		12	18.8	+46	06	14.8	IRR		265	
7412	N4298	12	19.0	+14	53	12.2	SC		500	
7414	N4299	12	19.1	+11	47	12.8	IRR		500	
7416		12	19.1	+41	08	14.2	SB SC			
7418	N4302	12	19.2	+14	53	13.4	SC		500	
7420	N4303	12	19.3	+04	45	10.9	SB-C SC		500	
7426	N4308	12	19.4	+30	20	14.3	E		263	
7429	N4319	12	19.5	+75	37	13.0	SB		269	
7431	N4307	12	19.6	+09	19	13.4	SB SA		500	
7435	N4309	12	19.7	+07	25	14.3	SO-A		500	
7440	N4310	12	19.9	+29	29	13.5	S... SO		263	
7442	N4312	12	20.0	+15	49	12.9	SA		500	
7443	N4314	12	20.0	+30	10	11.5	SA SC		263	
7445	N4313	12	20.1	+12	05	13.2	SA-B		500	
7446	N4318	12	20.2	+08	29	14.1	E		500	
7450	N4321	12	20.4	+16	06	10.6	SC		500	
7451	N4324	12	20.5	+05	32	12.5	SO SA		500	
7453	N4332	12	20.5	+66	07	13.2	SA		255	
7454	N4326	12	20.6	+06	21	13.1	S...			7461

UGC	V_r (km s^{-1})	Spectrum	Disp. (A/mm)	Ref.	Notes
7389					
7390	2602 MULT VEL	A	401-500	MW	
	1071	A	201-300	MW	
	1153	E	301-400	P	
7395	6697	N	NO INFO	CR 4	
7397	MULT VEL				
	1785	A	401-500	MW	
	1903	A	401-500	L	
7399	534	N	21-CM	FT	
7402	2736	E	101-200	SN 8	
7405	MULT VEL				
	750	A	401-500	L	
	948	N	21-CM	DL 1	
7407	390	E	301-400	LB	
7408	464	N	21-CM	FT	
7412	MULT VEL				
	1106	E	101-200	FR 8	
	1143	E	101-200	SN 8	
7414	MULT VEL				
	187	E	301-400	LB	
	310	N	101-200	EA	
7416	6927	E	101-200	FR11	
7418	MULT VEL				
	1309	E	1-100	SN 8	
	1123	N	101-200	EA	
7420	MULT VEL				
	1671	A	401-500	MW	
	1560	E	1-100	WD 1	
	1561	N	21-CM	DU	
	1560	N	21-CM	DL 1	
7426	606	A	201-300	CR 3	
7429	1700	N	NO INFO	A 7	
7431	1275	A	101-200	SN 8	
7435	1103	N	101-200	EA	
7440	901	A	201-300	CR 3	
7442	116	N	101-200	EA	
7443	883	A	401-500	MW	
7445	1579	N	101-200	EA	
7446	-300	E	201-300	AD 8	
7450	MULT VEL				
	1617	A	401-500	MW	
	1645	N	101-200	KT 1	
	1663	N	21-CM	DL 1	
	1537	N	21-CM	BG 3	
7451	MULT VEL				
	1714	A	401-500	MW	
	1781	N	101-200	EA	
7453	2843	N	101-200	KI	
7454	7286	N	101-200	EA	

UGC	NGC	R.A. (1950) DEC	m_{pg}	Hubble Type	Char.	Cluster	Group
7458	N4334	12 20.8 +07 45	14.9	SO			
7461	N4339	12 21.0 +06 21	13.1	E		500 BR IN GRP	7461
7463	N4346	12 21.0 +47 16	12.3	SO			
7465	N4343	12 21.1 +07 14	13.5	SB		500	
7466	N4342	12 21.1 +07 2C	13.0	S... SO		500	
7467	N4340	12 21.1 +17 00	12.4	SO		500	
7469	I3259	12 21.2 +07 28	14.7	...		500	
7470	I3258	12 21.2 +12 45	14.3	IRR		500	
7472	N4341	12 21.4 +07 23	14.5	SO		500	
7473	N4350	12 21.4 +16 58	11.5	SO		500	
7474	I3267	12 21.5 +07 19	14.6	S...		500	
7476	N4351	12 21.5 +12 29	13.5	S...		500	
7477	I3268	12 21.6 +06 53	14.2	PEC		500	
7483	N4359	12 21.7 +31 47	13.9	S...		263	
7484	N4360	12 21.8 +09 34	13.9	E			
7488	N4365	12 21.9 +07 35	11.5	E		500	
7489	N4369	12 22.1 +39 40	12.3	SO-A	MARK		
7490		12 22.1 +70 37	14.6	SC SC-I			
7491	N4386	12 22.1 +75 49	12.6	SO E		269	
7492	N4370	12 22.4 +07 43	14.1	SA		500	
7493	N4371	12 22.4 +11 59	12.1	SO		500	
7494	N4374	12 22.5 +13 10	10.8	SO E		500 BR IN GRP	7494
7496	N4375	12 22.5 +28 50	13.9	SA-B SO		263	
7497	N4378	12 22.7 +05 12	13.2	SA		500	
7501	N4377	12 22.7 +15 02	12.5	SO		500	
7502	N4379	12 22.7 +15 54	12.6	SO E		500	

102

UGC	V_r (km s^{-1})	Spectrum	Disp. (A/mm)	Ref.	Notes
7458					
	4377	N	101–200	EA	
7461					
	1278	A	401–500	MW	
7463					
	819	A	101–200	SN 8	
7465					
	1251	N	101–200	EA	
7466					
	714	A	401–500	MW	
7467	MULT VEL				
	824	A	101–200	SN 8	
	974	N	101–200	EA	
7469					
	1575	N	101–200	EA	
7470	MULT VEL				
	-408	E	301–400	UL 8	
	-517	N	1–100	FR12	
	-518	N	101–200	EA	
7472					
	1102	N	101–200	EA	
7473					
	1184	A	401–500	MW	
7474					
	2130	N	101–200	EA	
7476					
	2388	E	201–300	CR 2	
7477					
	764	E	201–300	CR 2	
7483					
	1171	N	NO INFO	CR 4	
7484	MULT VEL				
	7014	N	101–200	EA	COMP VEL A
	6676	N	101–200	EA	COMP VEL B
7488	MULT VEL				
	1171	A	401–500	MW	
	1290	A	401–500	L	
	1373	N	101–200	EA	
7489	MULT VEL				
	1020	E	101–200	DS 3	
	900	E	201–300	AD 7	
	1075	E	201–300	CR 2	
	981	N	NO INFO	HJ	
	1051	E	101–200	SN 8	
7490					
	470	N	21-CM	FT	
7491					
	1811	A	401–500	L	
7492					
	468	N	101–200	EA	
7493					
	977	A	301–400	V 5	
7494	MULT VEL				3C272.1
	1119	A	1–100	DC	
	954	E	401–500	MW	
	997	E	NO INFO	BB38	
	910	N	21-CM	DL 1	
7496	MULT VEL				
	9042	A	201–300	CR 3	
	9165	N	NO INFO	HG	
7497					
	2480	A	101–200	SN 8	
7501	MULT VEL				MAG SHARED
	1329	A	301–400	SA 2	
	1295	A	101–200	SN 8	
7502	MULT VEL				
	1009	A	101–200	SN 8	
	979	N	101–200	EA	

UGC	NGC	R.A. (1950) DEC	m$_{pg}$	Hubble Type	Char.	Cluster	Group
7503	N4380	12 22.8 +10 17	13.4	SA SA-B		500	
7506	N4384	12 22.8 +54 47	13.5	SA	MARK		
7507	N4383	12 22.9 +16 45	12.3	PEC SO		500	
7508	N4382	12 22.9 +18 28	10.2	SO		500	
7511	N4391	12 23.0 +65 13	13.8	SO		255	
7513		12 23.1 +07 30	14.4	SC		500	
7514	N4389	12 23.1 +45 58	12.8	S... PEC		265	
7515	N4385	12 23.2 +00 52	12.4	SO	MARK	500	
7517	N4387	12 23.2 +13 05	13.2	SA E		500	
7518	I3322	12 23.3 +07 50	14.7	SC-I		500	
7520	N4388	12 23.3 +12 56	12.2	SB		500	
7521	N4393	12 23.3 +27 50	13.8	SA-B SC		263	
7523	N4394	12 23.4 +18 30	11.9	SB		500	
7524	N4395	12 23.4 +33 49	11.7	... SC			
7528	N4402	12 23.6 +13 23	13.6	S...		500	
7529	N4405	12 23.6 +16 28	12.9	SO-A		500	
7532	N4406	12 23.7 +13 14	10.9	E		500	7494
7534		12 23.8 +58 35	15.3	IRR			
7535	N4410	12 23.9 +09 17	13.6	DBLE S...		266	
7536	N4412	12 24.0 +04 14	13.2	SC SB-C		500	
7537	N4411	12 24.0 +09 09	14.4	SC		266	
7538	N4413	12 24.0 +12 53	13.6	SA SA-B		500	
7539	N4414	12 24.0 +31 30	10.9	SC		267	
7540	N4415	12 24.1 +08 42	14.2	SA SO-A		500	
7542	N4417	12 24.2 +09 51	12.2	SO E-SO		500	
7546	N4411	12 24.3 +09 09	14.4	SC		266	
7549	N4420	12 24.4 +02 46	12.7	SC		500	

UGC	V_r (km s^{-1})	Spectrum	Disp. (A/mm)	Ref.	Notes
503	MULT VEL				
	970	A	101-200	SN 8	
	843	N	101-200	E A	
506					
	2400	E	201-300	A D 3	
507					
	1579	E	201-300	SN 8	
508					
	773	A	201-300	M W	
511					
	1337	N	101-200	K I	
513					
	984	N	21-CM	F T	
514					
	722	E	101-200	SN 8	
515	MULT VEL				
	1860	E	101-200	WK 2	
	2160	E	1-100	WD 1	
	2300	E	201-300	A S	
	2173	E	201-300	SN 8	
517					
	511	A	401-500	M W	
518					
	1177	N	101-200	E A	
520	MULT VEL				
	2604	E	101-200	FR 8	
	2420	E	201-300	SN 8	
521					
	842	N	NO INFO	CR 4	
523					
	772	A	401-500	M W	
524	MULT VEL				
	294	E	401-500	L	
	311	N	NO INFO	CR 4	
528					
	-11	N	101-200	E A	
529					
	1764	N	101-200	E A	
532	MULT VEL				
	-309	A	401-500	L	
	-374	A	401-500	M W	
	-100	A	301-400	L B	
	-299	N	101-200	E A	
534					
	724	N	21-CM	F T	
535	MULT VEL				
	7225	E	201-300	B C	
	7456	N	101-200	E A	COMP VEL A
	7567	N	101-200	E A	COMP VEL B
536					
	2271	E	101-200	SN 8	
537					
	-109	N	101-200	E A	
538	MULT VEL				
	-90	E	201-300	SN 8	
	-9	N	101-200	E A	
539					
	715	A	401-500	M W	
540					
	496	N	101-200	E A	
542	MULT VEL				
	796	A	101-200	SN 8	
	920	N	101-200	E A	
546					
	842	N	101-200	E A	
549	MULT VEL				
	1694	E	101-200	SN 8	
	1849	N	101-200	E A	

UGC	NGC	R.A. (1950) DEC	m_{pg}	Hubble Type	Char.	Cluster	Group
7551	N4419	12 24.4 +15 19	11.6	SA SA-B		500	
7554	N4421	12 24.5 +15 45	12.9	SO-A SA		500	
7559		12 24.6 +37 25	15.6	IRR			
7560		12 24.6 +48 33	14.7	S...	MARK		
7561	N4424	12 24.7 +09 42	13.1	S... SB SB		500	
7562	N4425	12 24.7 +13 01	13.3	SO-A SA		500	
7566	N4430	12 24.9 +06 32	13.4	SB	DISR	500	
7568	N4429	12 24.9 +11 23	11.4	SO		500	
7569	N4431	12 24.9 +12 34	14.5	SO		500	7581
7571	N4434	12 25.0 +08 26	13.2	E		500	
7572	N4441	12 25.0 +65 05	13.5	PEC		255	
7574	N4438	12 25.2 +13 17	12.0	S... SA	DISR	500	7494
7575	N4435	12 25.2 +13 21	11.9	SO	DISR	500	7494
7577		12 25.2 +43 46	14.7	IRR			
7578	I3376	12 25.3 +27 16	14.4	SA		263	
7581	N4440	12 25.4 +12 34	13.0	SA		500 BR IN GRP	7581
7583	N4442	12 25.5 +10 05	11.2	SO		500	
7587	N4445	12 25.7 +09 43	13.7	S...		500	
7591	N4448	12 25.8 +28 54	11.9	SA SB		263	
7592	N4449	12 25.8 +44 22	10.0	IRR			
7593		12 25.8 +44 43	14.8	DBLE	MARK		
7594	N4450	12 25.9 +17 21	11.2	SB		500	
7599		12 26.0 +37 30	15.6	SC-I			
7600	N4451	12 26.1 +09 32	13.4	...		500	
7601	N4452	12 26.2 +12 02	13.1	SO		500	

NGC	V_r (km s^{-1})	Spectrum	Disp. (A/mm)	Ref.	Notes
551	MULT VEL				
	-304	A	101-200	SN 8	
	-164	N	101-200	EA	
554					
	1692	A	701-1000	MW	
559					
	222	N	21-CM	FT	
560	MULT VEL				
	7800	E	201-300	AD 3	
	4627	N	101-200	SA 4	
					DISCREPANT VELOCITIES?
561	MULT VEL				
	452	E	201-300	BC	
	432	E	101-200	SN 8	
	599	N	101-200	EA	
562					
	1883	A	401-500	MW	
566					
	1436	N	101-200	EA	
568					
	1114	A	401-500	MW	
569					
	456	N	101-200	EA	
571					
	1007	N	101-200	EA	
572					
	1439	N	101-200	KI	
574	MULT VEL				
	-32	A	401-500	MW	
	321	E	301-400	BB38	
	31	N	101-200	EA	
575					
	869	A	401-500	MW	
577	MULT VEL				
	198	N	21-CM	FT	
	197	N	21-CM	BG 9	
578					
	7093	N	NO INFO	CR 4	
581					
	707	N	101-200	EA	
583					
	580	A	401-500	MW	
587					
	254	N	101-200	EA	
591					
	693	E	401-500	MW	
592	MULT VEL				
	309	E	301-400	LB	
	206	E	401-500	MW	
	200	N	NO INFO	SB	
	204	E	1-100	CT	
	216	E	1-100	V 6	
	190	N	21-CM	RR	
	204	N	21-CM	GU	
593	MULT VEL				
	6900	E	201-300	AD 4	
	7084	E	301-400	KR	COMP VEL A
	6834	E	301-400	KR	COMP VEL B
594					
	2048	A	401-500	MW	
599					
	280	N	21-CM	FT	
600	MULT VEL				
	600	E	201-300	AD 8	
	977	N	101-200	EA	
601	MULT VEL				
	182	A	101-200	SN 8	
	62	N	101-200	EA	

UGC	NGC	R.A. (1950) DEC	m_{pg}	Hubble Type	Char.	Cluster	Group
7603	N4455	12 26.2 +23 06	13.0	SC-I			
7606	N4454	12 26.3 −01 39	13.5	SC SO-A SA			
7608		12 26.3 +43 31	16.0	IRR			
7609	N4457	12 26.4 +03 51	11.9	SO-A		500	
7610	N4458	12 26.4 +13 31	13.3	E		500	7494
7611	N4460	12 26.4 +45 09	12.5	SO SB-C		265	
7612		12 26.5 +03 00	15.2	SC-I		500	
7613	N4461	12 26.5 +13 28	12.2	SO		500	7494
7614	N4459	12 26.5 +14 15	11.6	SO		500	
7615	I3407	12 26.5 +28 04	14.7	S...	DISR	263	
7619	N4464	12 26.8 +08 26	13.5	S... E		500	
7622	N4469	12 26.9 +09 01	12.6	SO-A SA		500	
7626	N4466	12 27.0 +07 58	14.7	S...		500	
7627	N4470	12 27.0 +08 06	12.9	S... PEC		500	
7629	N4472	12 27.2 +08 16	10.2	E		500 BR IN CLUS	
7631	N4473	12 27.3 +13 42	11.2	E		500	7494
7632	N4475	12 27.3 +27 32	14.6	SB SC		263	
7634	N4474	12 27.4 +14 21	12.6	SO		500	
7637	N4476	12 27.5 +12 37	13.3	SO E		500	
7638	N4477	12 27.5 +13 55	11.9	SO		500	7494
7645	N4478	12 27.8 +12 36	12.2	E		500	
7646	N4479	12 27.8 +13 51	13.9	SO		500	
7648	N4485	12 28.0 +41 59	12.4	IRR			
7649	N4483	12 28.1 +09 17	13.4	SO		500	
7651	N4490	12 28.1 +41 55	10.1	... SC	DISR		
7654	N4486	12 28.3 +12 40	10.4	E		500	

UGC	V_r (km s^{-1})	Spectrum	Disp. (A/mm)	Ref.	Notes
7603	MULT VEL				
	621	N	NO INFO	GT	
	601	E	101-200	SN 8	
7606	MULT VEL				
	2328	A	101-200	SN 8	
	2450	N	101-200	EA	
7608	543	N	21-CM	FT	MAG NOT ZW
7609	738	A	301-400	V 5	
7610	383	A	701-1000	MW	
7611	528	E	101-200	SN 8	
7612	1571	N	21-CM	FT	
7613	1887	A	401-500	MW	
7614	1111	A	401-500	MW	
7615	7072	N	NO INFO	CR 4	
7619	1199	A	401-500	MW	
7622	MULT VEL				
	468	A	101-200	SN 8	
	784	N	101-200	EA	
7626	1011	N	101-200	EA	
7627	MULT VEL				
	2307	E	201-300	SN 8	
	2441	N	101-200	EA	
7629	MULT VEL				
	926	A	101-200	DC	
	1013	A	201-300	MW	
	876	A	301-400	LB	
	1034	A	301-400	V 7	
	850	N	NO INFO	SB	
7631	MULT VEL				
	2241	A	201-300	MW	
	2281	A	1-100	MC	
7632	7349	E	1-100	FR11	
7634	1526	A	401-500	MW	
7637	1948	A	101-200	SN 8	
7638	1263	A	401-500	MW	
7645	1482	A	401-500	MW	
7646	822	A	401-500	MW	
7648	786	E	301-400	P	
7649	845	A	101-200	SN 8	
7651	MULT VEL				
	631	E	301-400	P	
	625	E	201-300	MW	
	476	A	301-400	LB	
7654	MULT VEL				3C274
	1225	E	1-100	DC	
	1290	E	201-300	MW	
	1196	E	401-500	L	
	1183	E	301-400	LB	
	1260	E	1-100	W	
	1297	A	201-300	J	

UGC	NGC	R.A. (1950) DEC				m_{pg}	Hubble Type	Char.	Cluster	Group
7655	N4489	12	28.3	+17	02	13.2	E		500	
7656	N4492	12	28.4	+08	21	14.1	S... SA		500	
7657	N4491	12	28.4	+11	46	13.7	...		500	
7658		12	28.4	+12	33	11.2	E COMP		500	
7662	N4494	12	28.9	+26	03	10.7	E			
7663	N4495	12	28.9	+29	25	14.1	SA-B		263	
7665	N4497	12	29.0	+11	54	13.8	SO		500	
7666	I3453	12	29.0	+15	08	15.2	IRR		500	
7667	N4500	12	29.0	+58	15	13.2	SA	MARK		
7668	N4496	12	29.1	+04	13	13.3	SC		500	
7669	N4498	12	29.1	+17	08	12.8	SC		500	
7673		12	29.3	+30	00	17.0	IRR		263	
7675	N4501	12	29.4	+14	42	10.6	SB-C SC		500	
7680	N4503	12	29.6	+11	27	12.4	SO		500	
7682	N4506	12	29.7	+13	42	14.2	S...		500	
7693	N4514	12	30.1	+30	00	14.2	SB-C		263	
7694	N4517	12	30.2	+00	24	12.4	SC		500	
7698		12	30.4	+31	49	15.6	IRR			
7706	N4521	12	30.6	+64	13	13.0	SO-A		255	
7709	N4519	12	31.0	+08	55	12.8	SC		500	
7711	N4522	12	31.1	+09	26	13.6	SC SB-C		500	
7713	N4523	12	31.3	+15	26	15.1	IRR		500	
7714	N4525	12	31.3	+30	34	13.0	SC SB			
7718	N4526	12	31.5	+07	58	10.6	SO		500	
7721	N4527	12	31.6	+02	56	12.4	SB		500	
7722	N4528	12	31.6	+11	36	12.9	SO		500	
7726	N4532	12	31.8	+06	45	12.3	IRR		500	
7727	N4535	12	31.8	+08	28	11.1	SC		500	

UGC	V_r (km s^{-1})	Spectrum	Disp. (A/mm)	Ref.	Notes
7655					
7656	863	N	101-200	E A	
7657	1735	A	401-500	M W	
7658	160	N	101-200	E A	
7662	81 MULT VEL	N	101-200	E A	
	1333	A	401-500	M W	
	1303	A	401-500	L	
	1308	A	1-100	M C	
7663	4364	N	NO INFO	CR 4	
7665	1342	N	101-200	E A	
7666	2546	E	101-200	FR 8	
7667	3000	E	201-300	AD 3	
7668	1773 MULT VEL	E	301-400	L B	
	1883	E	301-400	V 5	
7669	1500 MULT VEL	N	21-CM	F T	
	1638	N	101-200	E A	
7673	639	N	21-CM	F T	MAG NOT ZW
7675	2120 MULT VEL	A	401-500	M W	
	2016	N	21-CM	DL 1	
7680	1387 MULT VEL	A	201-300	SN 8	
	1348	N	101-200	E A	
7682	680	N	101-200	E A	
7693	8011	N	NO INFO	CR 4	
7694	1218	E	401-500	L	
7698	335	N	21-CM	F T	
7706	2426	N	101-200	K I	
7709	1213	E	401-500	L	
7711	2374 MULT VEL	E	101-200	SN 8	
	2508	N	101-200	E A	
7713	262	N	21-CM	F T	
7714	1131	E	201-300	CR 3	
7718	447 MULT VEL	A	401-500	M W	
	448	N	21-CM	DL 1	
7721	1727	A	201-300	M W	
7722	1337	N	101-200	E A	
7726	2154	E	301-400	V 5	
7727	1930 MULT VEL	E	101-200	M W	
	2097	E	401-500	L	
	1920	E	1-100	WD 1	
	1942	N	21-CM	DL 1	

UGC	NGC	R.A. (1950) DEC	m_{pg}	Hubble Type	Char.	Cluster	Group
7729	N4531	12 31.8 +13 21	13.3	SO		500	
7732	N4536	12 32.0 +02 28	12.3	SB-C SB		500	
7737	I3522	12 32.1 +15 29	17.0	IRR		500	
7739		12 32.2 +06 34	15.3	IRR		500	
7742	N4540	12 32.3 +15 50	12.5	...		500	
7747	N4545	12 32.4 +63 48	13.1	SC			
7752		12 32.9 -02 03	15.4	S...			
7753	N4548	12 32.9 +14 46	11.5	SB		500	
7754		12 32.9 +29 47	14.9	SB		263	
7757	N4550	12 33.0 +12 30	12.5	SO E		500	
7758	N4562	12 33.0 +26 08	14.6	SC			
7759	N4551	12 33.1 +12 32	13.1	E		500	
7760	N4552	12 33.1 +12 50	11.1	E		500	
7762	N4555	12 33.2 +26 48	13.5	E		263	
7765	N4556	12 33.3 +27 12	14.4	E-SO E E		263	
7766	N4559	12 33.4 +28 14	10.7	SC SB		263	
7768	N4561	12 33.6 +19 36	12.7	SC SC-I			
7769	N4566	12 33.6 +54 30	13.9	S...			
7772	N4565	12 33.8 +26 15	10.3	SB			
7773	N4564	12 33.9 +11 43	12.2	E		500	
7776	N4568	12 34.0 +11 31	12.5	SC SB		500	
7777	N4567	12 34.0 +11 32	12.5	SC SB		500	
7778	I3582	12 34.0 +26 28	14.3	COMP E			
7781	I3576	12 34.1 +06 54	15.2	SC-I		500	

UGC	V_r (km s^{-1})	Spectrum	Disp. (A/mm)	Ref.	Notes
7729					
7732	338	N	101-200	E A	
7737	1927	E	401-500	L	MAG NOT ZW
7739	661	N	21-CM	F T	
7742	2026	N	21-CM	F T	
7747	1238	N	101-200	E A	
7752	2703	N	101-200	K I	
7753	6000 MULT VEL	N	VERY LOW	C E	
	433	E	401-500	M W	
	495	N	21-CM	DL 1	
7754	MULT VEL				
	4752	E	201-300	TG 2	
	4653	N	NO INFO	CR 4	
7757	350	E	401-500	M W	
7758	MULT VEL				
	1379	E	201-300	TG 2	
	1316	N	NO INFO	CR 4	
7759	978	A	701-1000	M W	
7760	MULT VEL				
	172	A	1-100	D C	
	276	A	401-500	M W	
	247	A	401-500	L	
	201	N	101-200	E A	
7762	6694	A	201-300	CR 3	
7765	MULT VEL				
	7987	A	201-300	CR 3 COMP VEL A
	7402	A	201-300	CR 3 COMP VEL B
				DISCREPANT VELOCITIES?	
7766	MULT VEL				
	856	E	401-500	L	
	791	N	21-CM	GU	
	816	N	21-CM	S 3	
	805	N	21-CM	K S	
7768	1444	E	101-200	SN 8	
7769	5290	A	201-300	CR 3	
7772	MULT VEL				
	1223	A	401-500	M W	
	1174	E	401-500	L	
	1100	N	NO INFO	SB	
	1205	N	21-CM	K S	
7773	1015	A	301-400	V 5	
7776	MULT VEL				
	2413	A	401-500	L	
	2223	E	301-400	P	
	2297	N	21-CM	DL 1 VEL CONTAM
7777	MULT VEL				
	2284	A	401-500	L	
	2175	E	301-400	P	
	2297	N	21-CM	DL 1 VEL CONTAM
7778	7122	A	201-300	CR 3	
7781	MULT VEL				
	1080	N	21-CM	LS	
	1076	N	21-CM	FT	
	1074	N	21-CM	BA 4	

UGC	NGC	R.A. (1950) DEC	m_{pg}	Hubble Type	Char.	Cluster	Group
7783	I3585	12 34.1 +27 07	15.0	SO E		263	
7785	N4570	12 34.3 +07 31	11.8	SO E		500	
7786	N4569	12 34.3 +13 26	11.8	SB		500	
7788	N4571	12 34.4 +14 29	13.6	SC		500	
7791	I3598	12 34.8 +28 30	15.0	SA-B		263	
7793	N4578	12 35.0 +09 50	12.9	SO E		500	
7794	N4580	12 35.2 +05 38	13.1	SA-B SB-C		500	
7796	N4579	12 35.2 +12 05	11.5	SB		500	
7797	N4589	12 35.4 +74 28	12.0	E		269	
7804	N4586	12 35.9 +04 35	13.5	SA		500	
7811		12 36.4 +32 16	14.6	S...			
7812		12 36.4 +32 23	14.0	S...			
7819	N4592	12 36.7 -00 15	12.6	SC-I		500	
7822	I3617	12 36.9 +08 14	14.8	IRR		500	
7826	N4595	12 37.3 +15 34	12.8	S... SC		500	
7828	N4596	12 37.4 +10 27	12.4	SO		500	
7831	N4605	12 37.7 +61 53	10.8	S... SC			
7835	I3651	12 38.3 +27 00	14.4	SO			
7836		12 38.4 +29 45	14.9	SC			
7839	N4606	12 38.5 +12 11	12.7	S...		500	
7841		12 38.7 +01 40	14.3	...			
7842	N4608	12 38.7 +10 25	12.6	SO		500	
7843	N4607	12 38.7 +12 09	14.7	S...		500	
7845		12 38.8 +28 08	15.6	SB-C			
7850	N4612	12 39.0 +07 35	12.9	SO		500	
7851	N4614	12 39.0 +26 18	14.2	SO-A			7852
7852	N4615	12 39.1 +26 20	13.8	SC	DISR		7852 BR IN GRP

114

UGC	V_r (km s^{-1})	Spectrum	Disp. (Å/mm)	Ref.	Notes
7783					
	7412	A	201–300	CR 3	
7785					
	1730	A	401–500	MW	
7786	MULT VEL				
	−330	E	1–100	WD 1	
	−296	E	101–200	RG	
	−310	E	101–200	BB43	
	−249	N	101–200	EA	
7788					
	332	N	101–200	EA	
7791					
	7626	N	NO INFO	CR 4	
7793					
	2282	A	401–500	MW	
7794	MULT VEL				
	1260	A	101–200	SN 8	
	1139	N	101–200	EA	
7796	MULT VEL				
	1752	A	401–500	MW	
	1808	N	21–CM	DL 1	
7797					
	1825	A	401–500	MW	
7804	MULT VEL				
	799	A	101–200	SN 8	
	801	N	101–200	EA	
7811					
	6959	N	NO INFO	CR 4	
7812					
	4361	N	NO INFO	CR 4	
7819					
	1000	N	101–200	EA	
7822					
	2115	N	21–CM	FT	
7826					
	630	A	101–200	SN 8	
7828	MULT VEL				
	1920	A	101–200	SN 8	
	1975	N	101–200	EA	
7831	MULT VEL				
	140	E	401–500	L	
	195	E	101–200	BE 2	
7835					
	4785	A	201–300	TG 2	
7836					
	9377	E	201–300	TG 2	
7839					
	1660	N	101–200	EA	
7841					
	9300	E	201–300	AD10	
7842	MULT VEL				
	1840	A	101–200	SN 8	
	1805	N	101–200	EA	
7843					
	2440	N	101–200	EA	
7845					
	7631	E	201–300	TG 2	
7850	MULT VEL				
	1212	N	21–CM	BI 1	
	1802	A	101–200	SN 8	
	1843	N	101–200	EA	
7851	MULT VEL				DISCREPANT VELOCITIES?
	4917	E	201–300	TG 2	
	4789	N	NO INFO	CR 4	
7852	MULT VEL				
	4795	E	201–300	TG 2	
	4687	N	NO INFO	CR 4	

115

UGC	NGC	R.A. (1950) DEC	m pg	Hubble Type	Char.	Cluster	Group
7853	N4618	12 39.1 +41 26	11.5	••• SC			
7858	N4621	12 39.5 +11 55	11.0	E		500	
7860	N4627	12 39.5 +32 51	13.3	E			
7861	N4625	12 39.5 +41 34	13.0	••• SC-I			
7862	N4623	12 39.6 +07 57	13.6	S0-A E		500	
7863	I3672	12 39.6 +12 01	15.1	E		500	
7864	I 810	12 39.6 +12 52	14.7	•••		500	
7865	N4631	12 39.7 +32 49	9.8	SC			
7866	I3687	12 39.7 +38 46	15.5	IRR			
7870	N4632	12 40.0 +00 10	12.6	SC		500	
7871	N4630	12 40.0 +04 14	13.4	IRR SC-I		500	
7876	N4635	12 40.2 +20 12	13.7	SC			
7878	N4636	12 40.3 +02 58	11.8	E		500	
7880	N4638	12 40.3 +11 43	12.2	S0 E		500	
7884	N4639	12 40.4 +13 32	12.4	SB		500	
7887	N4644	12 40.4 +55 25	14.8	SB	DISR		7946
7888	N4640	12 40.5 +12 34	15.2	S•••		500	
7889	N4641	12 40.6 +12 19	14.9	S0		500	
7890		12 40.6 +28 00	14.5	•••			
7892	N4646	12 40.6 +55 07	13.8	•••			7946
7895	N4643	12 40.8 +02 16	11.9	S0		500	
7896	N4647	12 41.0 +11 52	12.5	SC	DISR	500	

UGC	V_r (km s^{-1})	Spectrum	Disp. (A/mm)	Ref.	Notes
7853	MULT VEL				
	562	E	301-400	V 5	
	573	E	301-400	KR	
	484	E	401-500	L	
7858					
	414	A	401-500	MW	
7860					
	727	N	NO INFO	CR 4	
7861	MULT VEL				
	847	E	301-400	KR	
	549	A	301-400	V 8	
	362	E	301-400	V 8	
7862					
	1933	A	101-200	SN 8	
7863					
	229	N	101-200	EA	
7864					
	-100	N	101-200	EA	
7865	MULT VEL				
	630	E	301-400	V 4	
	591	E	401-500	MW	
	665	E	301-400	LB	
	692	N	301-400	DA 2	
	600	N	21-CM	RR	
	600	N	21-CM	WE 1	
	630	N	21-CM	R 2	
	624	N	21-CM	KS	
	610	N	21-CM	GW 4	
7866	MULT VEL				
	352	N	21-CM	FT	
	367	N	21-CM	BG 3	
7870					
	1688	E	301-400	V 5	
7871					
	662	E	101-200	SN 8	
7876					
	992	E	101-200	SN 8	
7878	MULT VEL				
	736	A	301-400	LB	
	954	A	401-500	L	
	973	E	401-500	MW	
	1000	E	NO INFO	BB38	
	1090	N	21-CM	KG 2	
	1100	N	21-CM	BG 7	
7880					MAG SHARED
	1080	A	401-500	MW	
7884	MULT VEL				
	953	E	101-200	FR 8	
	725	A	101-200	SN 8	
7887	MULT VEL				
	4913	A	101-200	DP 3	COMP VEL A
	4764	A	101-200	DP 3	COMP VEL B
7888					MAG SHARED
	2087	N	101-200	EA	
7889					
	2305	N	101-200	EA	
7890					
	7456	E	201-300	TG 2	
7892					
	4551	A	101-200	DP 3	
7895					
	1432	A	401-500	L	
7896	MULT VEL				
	1073	E	301-400	P	
	1448	E	401-500	L	

UGC	NGC	R.A. (1950) DEC	m_{pg}	Hubble Type	Char.	Cluster	Group
7898	N4649	12 41.1 +11 50	10.3	E		500	
7901	N4651	12 41.3 +16 40	11.3	SC		500	
7902	N4654	12 41.4 +13 25	11.8	SC SC-I	DISR	500	
7905		12 41.5 +55 10	14.1	DBLE	MARK		7946
7907	N4656	12 41.6 +32 27	10.6	••• IRR			
7911		12 41.9 +00 45	14.9	SC-I		500	
7914	N4660	12 42.0 +11 28	12.1	E		500	
7915	N4659	12 42.0 +13 47	13.3	S0-A		500	
7916		12 42.0 +34 40	17.0	IRR			
7917	N4662	12 42.1 +37 23	14.1	SB SC			
7920	I3718	12 42.3 +12 37	14.7	S•••		500	
7922		12 42.3 +56 25	15.3	SB-C			
7924	N4665	12 42.5 +03 20	12.4	S0 SA		500	
7925	N4669	12 42.5 +55 09	15.1	S•••			7946
7926	N4666	12 42.6 −00 10	12.0	SC		500	
7928	I3734	12 42.7 +23 19	14.4	S•••	DISR		
7930	N4670	12 42.8 +27 24	12.6	PEC			
7932	I3742	12 43.0 +13 36	14.6	S•••		500	
7933	N4673	12 43.1 +27 20	13.7	S0			
7935	N4675	12 43.3 +55 00	15.4	SB			7946
7938	N4676	12 43.7 +31 00	14.1	DBLE S0	DISR		

UGC	V_r (km s^{-1})	Spectrum	Disp. (A/mm)	Ref.	Notes
7898	MULT VEL				
	1193	A	301-400	L B	
	1389	A	201-300	M W	
	1244	A	401-500	L	
	961	A	301-400	P	
	1090	N	NO INFO	S B	
7901	MULT VEL				
	733	A	301-400	V 5	
	750	E	NO INFO	BB45	
	796	N	21-CM	DL 1	
	810	N	21-CM	GS	
7902	MULT VEL				
	1042	E	101-200	FR 8	
	1022	E	301-400	V 5	
	1027	N	21-CM	DL 1	
	1040	N	21-CM	FR 8	
	1037	N	21-CM	S 3	
7905	MULT VEL				
	4912	E	101-200	DP 3	
	4904	E	301-400	SA 2	
	4843	E	1-100	CA	
	4920	E	NO INFO	DS 7	
7907	MULT VEL				
	721	E	401-500	L	
	827	N	NO INFO	DA 1	
	627	N	21-CM	GU	
	630	N	21-CM	WE 1	
	600	N	21-CM	R 2	
	649	N	21-CM	KS	
	650	N	21-CM	GW 4	
7911					
	1184	N	21-CM	FT	
7914					
	1017	A	201-300	MW	
7915					
	380	N	101-200	EA	
7916					MAG NOT ZW
	612	N	21-CM	FT	
7917					
	6869	N	101-200	FR11	
7920					
	953	N	101-200	EA	
7922					
	4805	E	201-300	CR 3	
7924					
	785	A	401-500	MW	
7925					
	4999	A	101-200	DP 3	
7926					
	1645	E	401-500	L	
7928					
	7049	E	201-300	TG 2	
7930	MULT VEL				
	1180	E	201-300	K	
	1113	E	201-300	CR 2	
	1216	N	401-500	RC	
	1130	E	101-200	SO 2	
7932					
	915	N	101-200	EA	
7933					
	6991	N	NO INFO	RC	
7935					
	4806	E	201-300	CR 3	
7938	MULT VEL				MAG SHARED
	6640	E	1-100	ST	
	6500	E	301-400	BB13	
	6564	E	101-200	TS	

119

UGC	NGC	R.A. (1950) DEC	m_{pg}	Hubble Type	Char.	Cluster	Group
7939	N4676	12 43.7 +31 00	14.1	DBLE SO-A	DISR		
7946	N4686	12 44.4 +54 49	13.7	SA			7946 BR IN GRP
7949		12 44.6 +36 45	17.0	IRR			
7950		12 44.6 +51 55	14.1	IRR			
7951	N4684	12 44.7 -02 27	12.4	SO			
7957	I 821	12 45.0 +30 04	14.5	SB-C			
7958	N4687	12 45.0 +35 37	14.3	SC ...	MARK		
7959		12 45.1 +27 16	14.7	SO		276	
7961	N4688	12 45.2 +04 36	14.5	SC		500	
7965	N4689	12 45.3 +14 02	12.8	SC		500	
7966	N4695	12 45.3 +54 39	14.5	S...			7946
7967	N4692	12 45.4 +27 30	14.0	E-SO		276	
7969	N4694	12 45.8 +11 15	12.4	SO PEC		500	
7970	N4698	12 45.9 +08 45	12.1	SA		500	
7971	N4707	12 46.1 +51 27	15.2	SC-I			
7972	N4704	12 46.3 +42 12	14.8	SB SC		273	
7975	N4701	12 46.7 +03 40	13.1	SC		500	
7977	N4712	12 47.1 +25 44	13.5	SC			
7978		12 47.1 +31 07	14.8	SC			
7980	N4710	12 47.2 +15 26	11.6	SO-A SO		500	
7985	N4713	12 47.4 +05 35	12.3	SC		500	
7986	N4715	12 47.4 +28 05	15.4	SO		276	
7987	N4719	12 47.7 +33 25	14.2	SB	MARK		
7989	N4725	12 48.0 +25 46	10.2	SB			
7994	N4750	12 48.2 +73 09	11.8	SB			
7996	N4736	12 48.5 +41 23	8.7	SB			

UGC	V_r (km s^{-1})	Spectrum	Disp. (A/mm)	Ref.	Notes
7939	MULT VEL				MAG SHARED
	6560	E	1-100	S T	
	6605	E	301-400	BB13	
7946					
	5015	A	101-200	DP 3	
7949					MAG NOT ZW
	333	N	21-CM	F T	
7950					
	322	E	101-200	DP 3	
7951	MULT VEL				
	1559	E	101-200	SN 8	
	1603	N	101-200	E A	
7957	MULT VEL				
	6579	A	201-300	TG 2	
	6850	A	201-300	K	
	6711	N	101-200	FR11	
7958					
	690	E	101-200	DS 3	
7959					
	7118	N	NO INFO	RC	
7961	MULT VEL				
	1000	N	NO INFO	ZG	
	981	N	21-CM	S 3	
7965	MULT VEL				
	1746	A	101-200	SN 8	
	1577	N	101-200	E A	
7966					
	4928	E	101-200	DP 3	
7967					
	7912	N	NO INFO	RC	
7969	MULT VEL				
	1234	A	101-200	SN 8	
	1146	N	101-200	E A	
7970	MULT VEL				
	1032	E	401-500	MW	
	872	N	21-CM	DL 1	
	1006	N	21-CM	KG 1	
7971					
	468	N	21-CM	F T	
7972					
	8098	E	101-200	FR11	
7975					
	750	N	101-200	E A	
7977	MULT VEL				
	4542	E	201-300	CR 3	
	4381	A	201-300	CR 2	
7978					
	8174	E	201-300	TG 2	
7980	MULT VEL				
	1125	E	301-400	V 5	
	1086	A	201-300	K	
7985					
	664	E	401-500	L	
7986					
	6897	A	201-300	K	
7987	MULT VEL				
	7105	E	101-200	DS 3	
	7170	E	201-300	TG 2	
7989	MULT VEL				
	1218	E	201-300	CR 2	
	1114	A	401-500	MW	
7994					
	1647	E	401-500	L	
7996	MULT VEL				
	296	E	301-400	BB20	
	282	E	401-500	MW	
	313	E	401-500	L	

UGC	NGC	R.A. (1950) DEC	m_{pg}	Hubble Type	Char.	Cluster	Group
7997	N4733	12 48.6 +11 11	13.2	E		500	
7999	N4738	12 48.7 +29 04	14.9	SC		276	
8005	N4747	12 49.3 +26 02	13.2	IRR			
8007	N4746	12 49.4 +12 21	13.3	S...		500	
8009	N4753	12 49.8 −00 56	11.7	... IRR		500	
8010	N4754	12 49.8 +11 35	11.6	S0		500	
8013		12 50.1 +27 01	15.7	S...		276	
8014	N4758	12 50.2 +16 07	14.1	IRR		500	
8016	N4762	12 50.4 +11 30	11.1	S0 SA		500	
8017		12 50.4 +28 39	14.5	...		276	
8018	N4765	12 50.7 +04 44	13.0	...		500	
8020	N4771	12 50.8 +01 32	13.3	SC−I SC		500	
8021	N4772	12 50.9 +02 26	12.9	SA		500	
8024		12 51.6 +27 25	14.9	IRR			
8025		12 51.6 +29 52	14.8	SB		276	
8026	N4787	12 51.7 +27 20	15.5	S0−A		276	
8028	N4789	12 51.9 +27 20	13.3	E−S0 E		276	
8033	N4793	12 52.2 +29 12	12.3	SC		276	
8034	N4810	12 52.3 +02 56	14.9	IRR		500	
8035	N4800	12 52.3 +46 48	12.0	SB			

UGC	V_r (km s^{-1})	Spectrum	Disp. (A/mm)	Ref.	Notes
	259	E	1-100	W	
	271	N	NO INFO	SB	
	262	N	101-200	DA 2	
	244	N	101-200	KT 2	
	240	N	21-CM	RR	
	290	N	21-CM	BY	
	307	N	21-CM	BV	
7997					
	1033	N	101-200	EA	
7999	MULT VEL				
	4791	A	201-300	TG 2	
	4575	A	201-300	K	
	4801	A	201-300	GY	
8005	MULT VEL				
	1219	E	201-300	CR 2	
	1200	N	21-CM	PS	
8007					
	1447	N	101-200	EA	
8009	MULT VEL				
	847	A	301-400	V 5	
	1364	A	401-500	L	
	1202	A	101-200	CH	
8010					
	1461	A	401-500	MW	
8013	MULT VEL				
	7981	A	201-300	GY	
	7171	N	NO INFO	ZS	
8014					DISCREPANT VELOCITIES?
	1233	N	21-CM	FT	
8016	MULT VEL				
	868	A	401-500	MW	
	885	A	301-400	LB	
	997	A	401-500	L	
	970	E	1-100	BP 2	
8017	MULT VEL				
	7012	E	201-300	TG 2	
	7103	A	201-300	K	
8018	MULT VEL				
	750	E	201-300	AD10	
	742	E	101-200	SN 8	
8020					
	1218	N	101-200	EA	
8021	MULT VEL				
	1057	A	101-200	SN 8	
	992	N	101-200	EA	
8024	MULT VEL				
	378	N	21-CM	FT	
	360	N	21-CM	BA 4	
8025	MULT VEL				
	6309	A	201-300	TG 2	
	6529	A	201-300	GY	
8026					
	7636	A	201-300	GY	
8028	MULT VEL				
	8372	A	401-500	L	
	8220	A	201-300	TG 2	
8033	MULT VEL				
	2529	E	401-500	L	
	2449	E	201-300	TG 2	
	2525	A	101-200	SN 8	
8034					
	899	E	301-400	P	
8035	MULT VEL				
	746	E	401-500	MW	
	747	N	NO INFO	HJ	

UGC	NGC	R.A. (1950) DEC	m_{pg}	Hubble Type	Char.	Cluster	Group
8037	N4795	12 52.5 +08 20	13.5	SO-A SA			
8038	N4798	12 52.5 +27 41	14.3	... E		276	
8049	N4807	12 53.1 +27 47	14.4	COMP		276	
8051	N4814	12 53.2 +58 36	12.4	SB			
8054	N4808	12 53.3 +04 34	12.5	SC SC-I		500	
8057	N4816	12 53.8 +28 01	14.8	E-SO		276	
8060	N4819	12 54.1 +27 15	14.1	SA		276	
8062	N4826	12 54.2 +21 57	8.9	SB			
8065	N4827	12 54.3 +27 26	14.1	SO		276	
8068		12 54.6 +48 34	14.4	PEC	DISR		
8069		12 54.8 +29 18	14.8	S...		276	
8070	N4839	12 55.0 +27 46	13.6	E		276	
8071		12 55.1 +28 28	15.4	S...		276	
8072	N4841	12 55.1 +28 45	13.5	E		276	
8073	N4841	12 55.1 +28 45	13.5	E		276	
8074		12 55.2 +02 58	15.4	SC-I		500	
8076		12 55.4 +29 55	15.2	SC		276	
8078	N4845	12 55.5 +01 50	12.9	SB SA		500	
8080		12 55.6 +27 08	15.2	SO-A		276	
8082	N4848	12 55.7 +28 31	14.2	SA-B		276	
8084		12 55.8 +03 04	15.0	SC-I		500	
8086	N4849	12 55.8 +26 40	14.5	SO		276	
8091		12 56.2 +14 29	15.3	IRR		500	

UGC	V_r (km s^{-1})	Spectrum	Disp. (A/mm)	Ref.	Notes
8037	MULT VEL				
	3169	A	201-300	SN 8	
	2714	N	101-200	EA	
					DISCREPANT VELOCITIES?
8038					
	7673	A	401-500	MW	
8049	MULT VEL				
	6973	A	201-300	TG 2	
	6846	A	201-300	K	
8051					
	2531	A	401-500	MW	
8054	MULT VEL				
	738	E	301-400	V 5	
	778	N	21-CM	S 3	
8057					
	6853	A	201-300	K	
8060					
	6696	N	NO INFO	RC	
8062	MULT VEL				
	382	E	201-300	MW	
	490	E	301-400	BB37	
	404	N	21-CM	DL 2	
	367	N	21-CM	BA 2	
8065	MULT VEL				
	7498	A	201-300	TG 2	
	7650	N	NO INFO	RC	
8068					
	8748	E	301-400	SA 2	
8069					
	7457	A	201-300	K	
8070	MULT VEL				5C04.51
	7440	A	201-300	TG 2	
	7446	N	NO INFO	RC	
8071					
	7078	A	201-300	GY	
8072	MULT VEL				MAG SHARED
	6879	A	201-300	TG 2	
	6695	A	201-300	K	VEL CONTAM
8073	MULT VEL				MAG SHARED
	6211	A	201-300	TG 2	
	6695	A	201-300	K	VEL CONTAM
					DISCREPANT VELOCITIES?
8074					
	926	N	21-CM	FT	
8076					
	5319	A	201-300	GY	
8078	MULT VEL				
	1198	A	101-200	SN 8	
	867	N	101-200	EA	
8080	MULT VEL				
	7371	A	201-300	GY	
	5700	A	NO INFO	F	
					DISCREPANT VELOCITIES?
8082	MULT VEL				
	7260	E	201-300	TG 2	
	7209	E	401-500	L	
8084					
	2725	N	21-CM	FT	
8086	MULT VEL				
	6028	A	201-300	TG 2	
	5829	N	401-500	RC	
8091	MULT VEL				
	257	E	NO INFO	HO	
	216	N	21-CM	FT	

UGC	NGC	R.A. (1950) DEC	m_{pg}	Hubble Type	Char.	Cluster	Group
8092	N4853	12 56.2 +27 51	14.2	COMP		276	
8096	I3949	12 56.5 +28 06	14.9	E / SB		276	
8097	N4859	12 56.6 +27 05	14.8	SO / SO-A		276	
8098	N4861	12 56.7 +35 08	12.8	DBLE / IRR	MARK		
8099	N4868	12 56.8 +37 34	12.9	SC-I / SA			
8100	N4865	12 56.9 +28 21	14.6	SC / E		276	
8102	N4866	12 57.0 +14 27	11.9	SA		500	
8103	N4874	12 57.2 +28 14	13.7	SO		276 2BR IN CLU	
8106	N4881	12 57.5 +28 31	14.7	E		276	
8108	N4892	12 57.6 +27 10	14.7	S...		276	
8109	N4880	12 57.7 +12 45	13.3	SO / E-SO		500	
8110	N4889	12 57.7 +28 15	13.0	E		276 BR IN CLUS	
8113	N4895	12 57.8 +28 28	14.3	SO		276	
8116	N4900	12 58.1 +02 46	12.8	SC		500	
8117	N4896	12 58.1 +28 35	15.1	E-SO / SO		276	
8118	I 842	12 58.2 +29 17	14.6	S...		276	
8121	N4904	12 58.4 +00 14	13.2	SC / SB-C		500	
8122		12 58.4 +27 40	15.5	SO-A		276	
8125	N4914	12 58.4 +37 35	12.7	E / SO			
8128	N4911	12 58.5 +28 04	13.7	S...	DISR	276	
8129	I4051	12 58.5 +28 17	14.8	SB / E		276	
8133	N4919	12 58.9 +28 04	14.9	SO		276	

UGC	V_r (km s^{-1})	Spectrum	Disp. (A/mm)	Ref.	Notes
8092					
	7550	A	201-300	MW	
8096	MULT VEL				
	7408	A	201-300	TG 2	
	7526	A	401-500	L	
8097					
	7030	A	201-300	K	
8098	MULT VEL				
	790	E	301-400	BB28	
	789	E	301-400	KR	COMP VEL A
	624	E	301-400	KR	COMP VEL B
	793	E	401-500	L	
	810	E	401-500	WK 2	
	828	E	1-100	CA	
	837	N	21-CM	CA	
	793	N	NO INFO	HJ	
8099					
	4701	E	101-200	SN 8	
8100	MULT VEL				
	4576	A	201-300	TG 2	
	4643	A	201-300	MW	
8102	MULT VEL				
	1910	E	201-300	MW	
	1986	N	21-CM	KG 1	
8103	MULT VEL				5C04.85
	7168	A	201-300	TG 2	
	7171	A	401-500	MW	
	7119	A	201-300	K	
	6959	A	201-300	CR 2	
8106	MULT VEL				
	6711	A	201-300	TG 2	
	6691	A	401-500	MW	
8108					
	5873	A	201-300	K	
8109	MULT VEL				
	1527	A	101-200	SN 8	
	1252	N	101-200	EA	
8110	MULT VEL				
	6481	A	201-300	TG 2	
	6416	A	401-500	MW	
	6585	A	401-500	L	
	6490	N	NO INFO	TI 2	
8113	MULT VEL				
	8387	A	201-300	TG 2	
	8406	A	401-500	MW	
8116					
	1054	E	401-500	L	
8117					
	5820	A	401-500	MW	
8118					
	7193	A	201-300	GY	
8121					
	1440	E	101-200	SN 8	
8122					
	6932	A	201-300	GY	
8125					
	4748	A	101-200	SN 8	
8128	MULT VEL				
	7819	A	201-300	TG 2	
	8006	A	401-500	L	
8129	MULT VEL				
	4913	A	201-300	TG 2	
	4932	A	401-500	MW	
8133	MULT VEL				
	7265	A	201-300	TG 2	
	7085	A	201-300	K	

UGC	NGC	R.A. (1950) DEC			m_{pg}	Hubble Type	Char.	Cluster	Group
8134	N4921	12 59.0	+28	08	13.7	S A		276	
8135	N4922	12 59.0	+29	35	14.2	DBLE		276	
8137	I 843	12 59.2	+29	24	14.8	S0		276	
8140	I4088	12 59.4	+29	19	14.8	SA-B		276	
8142	N4926	12 59.5	+27	53	14.1	E-S0		276	
8154	N4931	13 00.6	+28	17	14.4	S0		276	
8160	N4934	13 00.9	+28	17	15.0	S...		276	
8161		13 01.0	+26	49	15.5	S...		276	
8167	N4944	13 01.5	+28	28	13.3	SA-B		276	
8175	N4952	13 02.6	+29	23	13.6	E		276	
8178	N4957	13 02.8	+27	50	14.2	E		276	
8185	N4961	13 03.3	+28	00	13.5	SC		276	
8188	I4182	13 03.5	+37	52	14.0	SC			
8194	N4966	13 03.9	+29	20	13.9	S...		276	
8201		13 04.6	+67	58	14.1	IRR			
8206		13 05.2	+27	45	15.7	S0-A		276	
8229		13 06.5	+28	27	14.3	SB		276	
8230	I 853	13 06.5	+53	02	15.0	SB SC		285	
8236	N4999	13 07.0	+01	57	13.5	SB			
8241	N5000	13 07.4	+29-	10	14.0	SB		276	
8256	N5005	13 08.6	+37	19	10.6	SB			
8259		13 08.7	+29	50	15.3	SA-B			8260
8260	N5004	13 08.7	+29	54	14.3	S0			8260 BR IN GRP
8263		13 08.8	+03	40	15.4	SB-C			
8264		13 09.0	+84	53	14.5	PEC			
8270	N5012	13 09.1	+23	12	13.6	SC		283	
8271	N5014	13 09.2	+36	32	13.5	S...	MARK	281	

128

UGC	V_r (km s^{-1})	Spectrum	Disp. (A/mm)	Ref.	Notes
8134	MULT VEL				
	5473	A	201-300	TG 2	
	5459	A	401-500	L	
8135	MULT VEL				
	7010	E	201-300	TG 2	COMP VEL A
	7232	A	201-300	TG 2	COMP VEL B
	7357	N	NO INFO	RC	
8137	MULT VEL				
	7398	A	201-300	TG 2	
	7500	A	201-300	K	
8140					
	7027	A	201-300	GY	
8142	MULT VEL				
	7888	A	201-300	K	
	7668	N	401-500	RC	
8154					
	5824	A	201-300	K	
8160					
	6104	A	201-300	GY	
8161					
	6729	A	201-300	GY	
8167					MAG SHARED
	6993	N	NO INFO	RC	
8175					
	5865	A	401-500	L	
8178					
	6981	A	201-300	K	
8185	MULT VEL				
	2576	E	201-300	TG 2	
	2568	N	401-500	RC	
8188	MULT VEL				
	206	N	NO INFO	HJ	
	280	E	NO INFO	Z 4	
8194					MAG SHARED
	7077	E	201-300	K	
8201					
	34	N	21-CM	FT	
8206					
	6667	A	201-300	GY	
8229					
	5986	A	201-300	GY	
8230					
	7152	N	21-CM	FR11	
8236					
	3075	A	101-200	SN 8	
8241	MULT VEL				
	5706	E	201-300	CR 2	
	5620	A	201-300	K	
8256	MULT VEL				
	1025	E	301-400	BB14	
	1013	A	401-500	MW	
	1041	E	401-500	L	
	1003	N	21-CM	DL 2	
8259					
	7163	A	201-300	K	
8260					
	6957	A	201-300	K	
8263					
	2990	E	301-400	ZG	
8264					
	4641	E	301-400	CA	
8270	MULT VEL				
	2566	E	201-300	TG 2	
	2785	A	101-200	SN 8	
8271	MULT VEL				
	1170	E	101-200	DS 3	
	900	E	201-300	AD 7	

129

UGC	NGC	R.A. (1950) DEC		m pg	Hubble Type	Char.	Cluster	Group
8273		13 09.5	+21 04	15.5	SO			
8279	N5016	13 09.7	+24 21	14.3	SB-C			
8290		13 10.2	+23 06	14.8	SC PEC SC-I		283	
8292	N5025	13 10.4	+32 04	14.6	SB			
8300	N5032	13 11.0	+28 04	13.6	SB			
8303		13 11.0	+36 28	14.7	IRR		281	
8307	N5033	13 11.2	+36 51	10.9	SC			
8308		13 11.2	+46 35	17.0	IRR			
8319	N5041	13 12.2	+30 58	14.2	SC		287	
8320		13 12.2	+46 11	14.0	IRR		282	
8323		13 12.5	+35 08	14.9	IRR	MARK	281	
8327		13 13.0	+44 40	14.9	DBLE	MARK	282	
8330	N5052	13 13.2	+29 55	14.6	SO-A		287	
8331		13 13.3	+47 45	15.6	IRR			
8333		13 13.5	+25 42	17.0	IRR			
8334	N5055	13 13.5	+42 17	9.7	SB		284	
8335		13 13.6	+62 23	14.4	DBLE	DISR	279	
8337	N5056	13 13.8	+31 12	13.6	SC		287	
8342	N5057	13 14.1	+31 17	14.6	SO		287	BR IN CLUS
8356	N5065	13 15.2	+31 20	14.3	SC		287	
8365		13 16.5	+42 12	15.5	SC			
8366	N5081	13 16.8	+28 46	14.3	SB			
8371	N5089	13 17.3	+30 31	14.4	S... PEC			
8375	I 881	13 17.5	+16 07	14.8	SA			
8385		13 18.2	+10 03	14.6	... SC-I			
8387	I 883	13 18.3	+34 25	14.8	PEC	DISR		
8397		13 19.3	+31 37	14.8	SB-C		287	

UGC	V_r (km s^{-1})	Spectrum	Disp. (A/mm)	Ref.	Notes
8273					4C21.39
	9000	A	301-400	UL 9	
8279	MULT VEL				
	2769	E	201-300	TG 2	
	2553	E	101-200	SN 8	
8290					
	2592	E	201-300	TG 2	
8292					
	6394	A	201-300	TG 2	
8300					
	6536	E	201-300	TG 2	
8303					
	956	N	21-CM	FT	
8307	MULT VEL				
	894	E	101-200	FR 8	
	924	A	201-300	MW	
	908	E	401-500	L	
	813	E	301-400	LB	
	913	N	21-CM	R 3	
8308					MAG NOT ZW
	165	N	21-CM	FT	
8319					
	7471	A	201-300	TG 2	
8320					
	198	N	21-CM	FT	
8323					
	840	E	101-200	DS 3	
8327	MULT VEL				
	10800	E	201-300	AD 4	
	10231	E	301-400	KR	COMP VEL A
	10579	E	301-400	KR	COMP VEL B
8330					
	6773	A	201-300	TG 2	
8331					
	258	N	21-CM	FT	
8333					MAG NOT ZW
	936	N	21-CM	FT	
8334	MULT VEL				
	560	E	301-400	BB 6	
	500	E	201-300	MW	
	538	E	401-500	L	
	450	N	NO INFO	SB	
	510	N	21-CM	RR	
8335	MULT VEL				
	9471	E	301-400	KR	COMP VEL A
	9241	E	301-400	KR	COMP VEL B
8337					
	5481	E	201-300	TG 2	
8342					
	5856	A	201-300	TG 2	
8356					
	5732	E	201-300	TG 2	
8365					
	1215	N	21-CM	FT	
8366					
	6731	A	201-300	TG 2	
8371					
	2190	E	201-300	TG 2	
8375					
	6843	N	101-200	TU	
8385					
	1133	N	21-CM	FT	
8387	MULT VEL				
	6947	E	301-400	SA 2	
	6870	E	101-200	BB35	
8397					
	5088	E	201-300	TG 2	

UGC	NGC	R.A. (1950) DEC	m pg	Hubble Type	Char.	Cluster	Group
8399		13 19.4 +31 30	14.9	SB		287	
8403	N5112	13 19.7 +39 00	12.5	SC			
8410	N5116	13 20.6 +27 15	13.7	SC			
8411	N5117	13 20.6 +28 35	14.5	SC			
8419	N5127	13 21.4 +31 50	13.9	E		287 2BR IN CLU	
8420	N5144	13 21.4 +70 47	13.2	S...	MARK		
8433	N5141	13 22.6 +36 38	13.9	SO			
8441		13 23.5 +58 05	15.6	IRR		294	
8443	N5147	13 23.8 +02 21	12.7	... SC-I			
8448		13 24.3 +20 13	14.9	SB-C SC			
8451		13 24.6 +32 28	14.6	SC		287	
8458	N5164	13 25.2 +55 45	14.6	SB	MARK	285	
8466		13 26.1 +31 05	15.5	SB-C		287	
8468	N5173	13 26.3 +46 51	13.5	E			
8477	N5172	13 26.9 +17 19	12.7	SC SB			
8489		13 27.6 +45 39	15.2	... SC-I			
8490	N5204	13 27.7 +58 40	11.7	... SC		294	8981
8493	N5194	13 27.8 +47 27	8.8	SC			
8494	N5195	13 27.9 +47 31	10.6	IRR E			
8499	N5198	13 28.1 +46 56	13.2	E			
8502		13 28.3 +31 32	14.6	DBLE	MARK		
8553	N5223	13 32.1 +34 57	14.4	E		BR IN GRP	8553
8555	I 900	13 32.2 +09 36	14.3	SC			
8561		13 32.6 +34 18	13.8	SC			

UGC	V_r (km s^{-1})	Spectrum	Disp. (A/mm)	Ref.	Notes
8399					
	7272	E	201-300	TG 2	
8403					
	965	N	21-CM	S 3	
8410					
	2589	E	201-300	TG 2	
8411					
	2465	E	201-300	TG 2	
8419	MULT VEL				B2 1321+31
	4830	A	NO INFO	UL17	
	4846	A	101-200	SN 8	
8420					
	3000	E	201-300	AD 3	
8433	MULT VEL				4C36.24
	5220	A	301-400	T	
	5250	A	101-200	SA 4	
8441					
	1519	N	21-CM	FT	
8443					
	1115	E	301-400	V 5	
8448					
	7144	E	101-200	FR11	
8451					
	5293	E	101-200	FR11	
8458					
	4800	E	201-300	AD 3	
8466					4C31.42
	7230	E	201-300	WL	
8468					
	2404	E	401-500	MW	
8477					
	4337	A	101-200	SN 8	
8489	MULT VEL				
	1297	N	21-CM	FT	
	1304	N	21-CM	BA 4	
8490	MULT VEL				
	272	E	401-500	L	
	208	N	21-CM	GU	
	205	N	21-CM	R 3	
	200	N	21-CM	AL 3	
8493	MULT VEL				
	438	E	201-300	MW	
	381	E	301-400	LB	
	472	E	301-400	P	
	474	E	1-100	BB34	
	468	E	1-100	CC	
	464	E	1-100	TL	
	510	E	1-100	WD	
	413	A	1-100	V 6	
	454	N	21-CM	DD	
	455	N	21-CM	RR	
	465	N	21-CM	WG	
8494	MULT VEL				
	542	E	201-300	MW	
	508	E	301-400	LB	
	562	E	301-400	P	
	606	E	1-100	BB34	
8499	MULT VEL				
	2482	A	401-500	MW	
	2562	A	401-500	L	
8502					
	10195	E	101-200	DS 3	
8553					
	7180	A	301-400	A 1	
8555					
	7080	E	101-200	FR11	
8561					
	7130	N	201-300	A 10	

UGC	NGC	R.A. (1950) DEC	m_pg	Hubble Type	Char.	Cluster	Group
8566	N5227	13 32.8 +01 40	14.6	SB			
8573	N5230	13 33.1 +13 56	13.4	SC		291 BR IN GRP	8573
8578		13 33.3 +29 29	15.1	...			
8588		13 33.7 +46 11	15.3	SC-I			
8597		13 34.2 +46 28	15.2	SC			
8614		13 34.9 +07 54	15.4	IRR			
8616	N5248	13 35.0 +09 08	11.4	SC			
8632	N5256	13 36.2 +48 32	14.1	PEC	MARK		
8641	N5257	13 37.4 +01 05	13.7	S... SB	DISR		
8645	N5258	13 37.5 +01 05	13.8	S... SB	DISR		
8651		13 37.8 +41 00	15.3	IRR			
8658		13 38.7 +54 35	14.4	SB-C SC			
8672	N5283	13 39.6 +67 55	14.3	S0	SEYF		
8675	N5273	13 39.8 +35 54	12.7	S0			
8677	N5278	13 39.8 +55 56	13.6	DBLE SB	MARK		
8678	N5279	13 39.8 +55 56	13.6	DBLE SA	MARK		
8680	N5276	13 40.1 +35 53	14.6	SA			
8683		13 40.4 +39 55	15.7	IRR			
8696		13 42.8 +56 08	15.0	PEC	SEYF		
8699	N5289	13 43.0 +41 45	13.5	SA-B			
8700	N5290	13 43.1 +41 58	13.0	SB-C			
8709	N5297	13 44.3 +44 07	12.3	SC			
8710	N5293	13 44.4 +16 31	14.3	SC			
8711	N5301	13 44.4 +46 22	13.0	SC			
8713		13 44.8 +34 09	15.5	SC			
8715		13 44.9 +34 08	14.8	SC S...			

UGC	V_r (km s^{-1})	Spectrum	Disp. (A/mm)	Ref.	Notes
8566					
	5198	E	101–200	FR11	
8573					
	6863	A	101–200	SN 8	
8578	MULT VEL				
	855	E	101–200	DP 3	
	838	N	21–CM	BG 2	
8588					
	1447	N	21–CM	FT	
8597					
	2427	N	21–CM	FT	
8614					
	1053	N	21–CM	FT	
8616	MULT VEL				
	1190	E	301–400	BB22	
	1176	A	401–500	MW	
	1232	A	401–500	L	
	1148	N	21–CM	DL 2	
	1125	N	21–CM	BA 1	
8632	MULT VEL				
	8400	E	201–300	AD 3	
	8235	E	201–300	BR 2	
8641	MULT VEL				
	6820	E	301–400	P	
	6706	N	101–200	TU	
8645	MULT VEL				
	6645	E	301–400	P	
	6638	N	101–200	TU	
8651					
	200	N	21–CM	FT	
8658					
	1977	E	101–200	FR11	
8672					
	2700	E	201–300	AD 3	
8675	MULT VEL				
	1022	E	201–300	MW	
	1021	N	101–200	TU	
	1032	N	21–CM	BI 1	
8677	MULT VEL				MAG SHARED
	7518	E	301–400	P	
	7800	E	201–300	AD 3	
	7570	N	101–200	TU	
8678	MULT VEL				MAG SHARED
	7561	E	301–400	P	
	7500	E	201–300	AD 3	
	7585	N	101–200	TU	
8680					
	5278	N	101–200	TU	
8683					
	663	N	21–CM	FT	
8696					
	11400	E	201–300	AD 3	
8699					
	2387	N	101–200	KI	
8700					
	2518	N	101–200	KI	
8709					
	2581	N	101–200	TU	
8710					
	5789	E	101–200	FR11	
8711	MULT VEL				
	1702	A	401–500	L	
	1705	N	NO INFO	HJ	
8713					
	4914	E	301–400	P	
8715					
	4450	E	301–400	P	

UGC	NGC	R.A. (1950) DEC	m_{pg}	Hubble Type	Char.	Cluster	Group
8718		13 45.1 +34 24	14.5	S...	MARK		
8722	N5308	13 45.4 +61 13	12.5	SO-A SO		294	
8725	N5303	13 45.6 +38 33	12.9	PEC		306	
8744	N5313	13 47.6 +40 14	12.4	S... SB		306	
8745	N5322	13 47.6 +60 26	11.3	E		294	
8760		13 48.6 +38 15	15.4	IRR		306	
8764	N5326	13 48.7 +39 49	12.9	SA SB		306	
8765	I 954	13 48.9 +71 25	14.5	PEC			
8782		13 50.0 +31 42	15.6	S... E		305	
8805	N5347	13 51.1 +33 45	13.3	SA-B SB		304	
8809	N5351	13 51.3 +38 10	13.1	SB		306	
8810	N5350	13 51.3 +40 36	12.4	SB-C		306	8813
8813	N5353	13 51.4 +40 31	11.8	SO E-SO		306 BR IN GRP	8813
8814	N5354	13 51.4 +40 32	12.3	SO E		306	8813
8819	N5355	13 51.6 +40 35	14.0	SO		306	8813
8823		13 51.8 +69 34	14.5	SO	SEYF	BR IN GRP	8823
8827		13 52.1 +15 17	14.1	SO			
8835	N5362	13 52.8 +41 33	13.2	PEC SB		306	
8837		13 52.9 +54 09	14.2	IRR			8981
8839		13 53.0 +18 01	15.7	IRR			
8843	N5372	13 53.1 +58 55	13.7	PEC		294	
8846	N5371	13 53.5 +40 42	11.5	SB		306	
8847	N5363	13 53.6 +05 30	11.4	... IRR		308	
8850		13 53.6 +18 37	14.8	PEC	SEYF		

136

UGC	V_r (km s^{-1})	Spectrum	Disp. (A/mm)	Ref.	Notes
8718	MULT VEL				
	4800	E	201-300	AD 7	
	4850	E	NO INFO	DS 6	
8722	MULT VEL				
	2046	A	401-500	MW	
	2035	A	401-500	L	
8725					
	1284	N	101-200	KI	
8744					
	2576	A	101-200	SN 8	
8745	MULT VEL				
	1902	A	401-500	MW	
	1629	N	101-200	KI	
8760	MULT VEL				
	189	N	21-CM	FT	
	206	N	21-CM	BA 4	
8764					
	2524	A	101-200	SN 8	
8765					
	8811	E	301-400	CA	
8782	MULT VEL				3C 293
	13500	E	301-400	BU	
	13463	E	301-400	SN 8	
8805					
	2266	E	101-200	SN 8	
8809					
	3845	A	101-200	SN 8	
8810	MULT VEL				
	2356	N	101-200	KI	
	2221	E	101-200	SN 8	
8813	MULT VEL				
	1990	E	301-400	P	
	2188	A	401-500	MW	
	2297	N	101-200	KI	
8814	MULT VEL				
	2980	A	301-400	P	
	2413	N	101-200	KI	DISCREPANT VELOCITIES?
8819					
	2368	N	101-200	KI	
8823	MULT VEL				
	9600	E	201-300	AD 3	
	9060	E	NO INFO	WD 2	DISCREPANT VELOCITIES?
8827					
	5700	E	201-300	AD 9	
8835					
	2232	A	101-200	SN 8	
8837	MULT VEL				
	149	E	401-500	L	
	141	N	21-CM	FT	
	139	N	21-CM	AL 3	
	96	E	101-200	SN 8	
8839					
	965	N	21-CM	FT	
8843					
	1711	N	101-200	KI	
8846	MULT VEL				
	2551	E	401-500	MW	
	2633	E	401-500	L	
8847	MULT VEL				
	1138	E	401-500	MW	
	1138	E	401-500	L	
8850					
	15140	E	101-200	DS 3	

UGC	NGC	R.A. (1950) DEC	m pg	Hubble Type	Char.	Cluster	Group
8852	N5376	13 53.6 +59 45	12.9	SA-B		294	
8853	N5364	13 53.7 +05 16	13.2	SB SB-C SC		308	
8863	N5377	13 54.3 +47 29	12.5	SA			
8865	N5375	13 54.5 +29 25	13.2	SB			
8866	N5389	13 54.5 +59 59	13.2	SO		294	
8869	N5378	13 54.7 +38 02	13.8	SA		306	
8870	N5380	13 54.8 +37 51	13.5	SO		306	
8875	N5383	13 55.0 +42 05	12.5	SB	MARK		
8877		13 55.1 +42 02	16.5	... SC-I			
8898	N5394	13 56.4 +37 42	13.7	SB	DISR	306	
8900	N5395	13 56.5 +37 40	12.6	SA-B SB		306	
8920		13 57.9 +13 12	15.5	SC	DISR	312	
8925	N5406	13 58.2 +39 09	13.1	SB		306	
8935	N5422	13 58.9 +55 24	13.1	SC SO-A			
8937	N5430	13 59.1 +59 33	12.7	SB		294	
8941	N5421	13 59.5 +34 04	14.3	DBLE	DISR		
8969	N5448	14 01.0 +49 25	12.7	SB SA			
8974	N5444	14 01.2 +35 23	12.8	SB-C E		BR IN GRP	8974
8981	N5457	14 01.4 +54 35	8.7	SC	DISR	BR IN GRP	8981
8994		14 02.3 -00 22	14.7	SB-C			
9002		14 02.4 +12 58	15.3	SC S...	DISR		
9011	N5473	14 03.0 +55 08	12.5	SO		316	

UGC	V_r (km s^{-1})	Spectrum	Disp. (A/mm)	Ref.	Notes
3852	MULT VEL				
	2077	N	101-200	K I	
	2034	A	101-200	SN 8	
3853					
	1393	A	401-500	M W	
3863					
	1830	E	401-500	M W	
3865					
	2083	A	301-400	V 8	
866					
	1835	N	101-200	K I	
3869					
	2967	N	101-200	K I	
3870	MULT VEL				
	2848	N	101-200	K I	
	3126	A	101-200	SN 8	
875	MULT VEL				
	2264	E	301-400	BB24	
	2100	E	201-300	AD 3	
	2264	E	101-200	FR13	
	2264	N	21-CM	FR13	
3877					MAG NOT ZW
	2379	E	101-200	FR13	
898	MULT VEL				
	3558	E	401-500	M W	
	3420	E	1-100	A 4	
	3282	E	301-400	K R	
	3325	E	101-200	SN 8	
900	MULT VEL				
	3459	N	21-CM	C X	
	3512	A	101-200	SN 8	
920					
	480	E	NO INFO	F	
925	MULT VEL				
	5151	N	101-200	K I	
	5248	A	101-200	SN 8	
935					
	1807	A	101-200	SN 8	
937	MULT VEL				
	2819	N	101-200	K I	
	3085	E	101-200	SN 8	
941					
	8049	E	301-400	K R	
969	MULT VEL				
	1970	E	401-500	M W	
	1973	N	NO INFO	H J	
974	MULT VEL				B2 1401+35
	3878	E	101-200	D C	
	3990	A	NO INFO	UL17	
	3964	A	101-200	SN 8	
981	MULT VEL				
	247	E	201-300	M W	
	160	E	301-400	L B	
	239	N	21-CM	D D	
	266	N	21-CM	D B	
	240	N	21-CM	R S	
	235	N	21-CM	R R	
	225	N	21-CM	GW 2	
994					
	7424	E	1-100	FR11	
002					
	4200	E	201-300	AD 9	
011	MULT VEL				
	1976	A	401-500	M W	
	2141	A	401-500	L	

UGC	NGC	R.A. (1950) DEC	m$_{pg}$	Hubble Type	Char.	Cluster	Group
9013	N5474	14 03.2 +53 54	11.9	SC SC-I	DISR		8981
9018	N5477	14 03.8 +54 42	14.5	SC-I IRR		316	
9026	N5480	14 04.5 +50 58	12.6	SC	DISR		
9029	N5481	14 04.9 +50 59	13.5	E E-SO			
9033	N5485	14 05.5 +55 15	12.4	SO		316	
9036	N5486	14 05.7 +55 20	14.0	SC-I SC		316	
9059	I 982	14 07.6 +17 56	14.6	SO		312	9058
9061	I 983	14 07.7 +17 58	14.3	SA-B		312	9058
9079	N5496	14 09.0 -00 55	13.4	SC			
9098		14 10.6 +45 55	14.1	DBLE S... S...			
9114	I 989	14 12.3 +03 21	14.4	E			9114 BR IN GRP
9119	N5523	14 12.6 +25 33	13.4	SC			
9122	N5521	14 12.9 +04 38	14.3	...			
9126		14 13.3 +16 46	16.0	IRR		312	
9128		14 13.6 +23 17	15.3	IRR			9128 BR IN GRP
9133	N5533	14 14.0 +35 34	13.0	SB		320	
9136	N5536	14 14.3 +39 44	14.5	SA			9139
9137	N5532	14 14.4 +11 02	13.3	SO E			
9139	N5541	14 14.4 +39 49	13.4	S... PEC			9139 BR IN GRP
9142	N5544	14 15.0 +36 48	13.2	SA SO		320	
9143	N5545	14 15.0 +36 48	13.2	SB-C		320	
9149	N5548	14 15.7 +25 22	13.1	SA	SEYF		
9161	N5557	14 16.3 +36 43	12.2	E			
9175	N5566	14 17.8 +04 09	12.0	SA			9175 BR IN GRP

UGC	V_r (km s^{-1})	Spectrum	Disp. (A/mm)	Ref.	Notes
9013	MULT VEL				
	247	E	401-500	L	
	204	E	1-100	SN 8	
	280	N	21-CM	GU	
	273	N	21-CM	S 3	
	275	N	21-CM	AL 3	
9018	MULT VEL				
	313	E	101-200	SN 8	
	312	N	21-CM	AL 3	
9026	MULT VEL				
	1787	E	301-400	P	
	2028	N	101-200	TU	
9029	MULT VEL				
	2092	A	301-400	P	
	2337	N	101-200	TU	
9033					
	1985	A	401-500	MW	
9036	MULT VEL				
	1317	E	101-200	SN 8	
	2845	N	21-CM	S 3	
					DISCREPANT VELOCITIES?
9059	MULT VEL				
	5053	A	301-400	A 1	
	0	N	NO INFO		VEL CONTAM
9061	MULT VEL				
	5053	A	301-400	A 1	
	0	N	NO INFO		VEL CONTAM
9079					
	1527	N	21-CM	BA 1	
9098	MULT VEL				
	8051	N	101-200	TU	COMP VEL A
	8273	N	101-200	TU	COMP VEL B
9114					
	7490	N	101-200	TU	
9119	MULT VEL				
	762	N	21-CM	LS	
	1047	N	21-CM	S 3	
9122					
	12300	E	201-300	AD10	
9126					MAG NOT ZW
	2277	N	21-CM	FT	
9128					
	153	N	21-CM	FT	
9133					
	3781	E	401-500	MW	
9136					
	5142	N	101-200	TU	
9137					MAG SHARED 3C296
	7087	A	301-400	SN 8	
9139	MULT VEL				
	7477	N	101-200	TU	
	6000	E	201-300	AD 8	
	1763	N	101-200	KI	
					DISCREPANT VELOCITIES?
9142	MULT VEL				MAG SHARED
	3172	A	301-400	P	
	3222	N	101-200	TU	
9143	MULT VEL				MAG SHARED
	3182	E	301-400	P	
	3261	N	101-200	TU	
9149	MULT VEL				
	4930	E	201-300	MW	
	5120	N	301-400	DA 2	
9161					
	3195	A	401-500	MW	
9175	MULT VEL				
	1643	A	301-400	LB	
	1455	A	401-500	MW	

UGC	NGC	R.A. (1950) DEC	m_{pg}	Hubble Type	Char.	Cluster	Group
9179	N5585	14 18.2 +56 57	11.7	SC		294	8981
9181	N5574	14 18.4 +03 28	13.4	SO			
9183	N5576	14 18.5 +03 30	12.3	E		BR IN GRP	9183
9189	N5607	14 18.7 +71 50	13.9	PEC	MARK		
9201	N5584	14 19.8 −00 09	12.8	SC			
9208	N5596	14 20.4 +37 20	14.5	SO	MARK		
9211		14 20.6 +45 37	15.7	IRR			
9214		14 20.8 +33 04	14.5	SA	SEYF		
9220	N5600	14 21.5 +14 52	11.9	S... SB			
9226	N5614	14 22.0 +35 05	12.6	SA			
9234		14 22.6 +26 22	15.7	S...	DISR	324	
9240		14 22.8 +44 45	13.9	IRR			
9241		14 23.2 +32 42	14.2	PEC			
9261	N5631	14 25.0 +56 49	12.4	SO−A SO			
9271	N5633	14 25.6 +46 22	12.9	SB COMP			
9283	N5635	14 26.3 +27 38	13.9	S...			
9300	N5641	14 27.0 +29 03	13.6	SB SA−B		326	
9308	N5638	14 27.2 +03 27	12.5	E		327 BR IN CLUS	
9318	N5653	14 27.9 +31 26	12.7	S... SC			
9324		14 28.0 +44 40	15.7	... SC−I			
9325	N5660	14 28.0 +49 50	12.2	SC			
9328	N5645	14 28.2 +07 30	12.8	... SC			
9347	N5673	14 29.8 +50 10	14.0	SC			
9352	N5665	14 30.0 +08 18	12.6	SC SC−I		330	
9353	N5669	14 30.3 +10 06	13.2	SC		330	
9354	N5672	14 30.4 +31 53	14.5	PEC SB			
9358	N5678	14 30.6 +58 09	12.1	SB	DISR		

UGC	V_r (km s^{-1})	Spectrum	Disp. (A/mm)	Ref.		Notes
9179	MULT VEL					
	304	E	401-500	L		
	307	N	21-CM	GU		
	298	N	21-CM	R	3	
	305	N	NO INFO	HJ		
	303	N	21-CM	AL	3	
9181						
	1716	A	401-500	MW		
9183						
	1528	A	401-500	MW		
9189						
	7800	E	201-300	AD	3	
9201						
	1651	E	201-300	V	7	
9208						
	4500	E	201-300	AD	7	
9211						
	690	N	21-CM	FT		
9214						
	10200	E	201-300	AD	7	
9220	MULT VEL					
	2700	E	201-300	AD	10	
	2333	E	101-200	SN	8	
9226						
	3872	A	401-500	MW		
9234						PKS1422+26
	10200	E	301-400	UL	9	
9240						
	153	N	21-CM	FT		
9241						
	3900	E	201-300	AD	8	
9261	MULT VEL					
	1979	E	401-500	MW		
	1950	N	21-CM	KG	1	
9271	MULT VEL					
	2316	E	401-500	MW		
	2390	A	401-500	L		
	2350	N	NO INFO	HJ		
	2332	N	21-CM	CX		
9283						
	3819	E	101-200	DP	3	
9300						
	4437	A	101-200	SN	8	
9308	MULT VEL					
	1677	A	401-500	MW		
	1733	N	101-200	TU		
9318						
	3557	E	401-500	L		
9324						
	2745	N	21-CM	FT		
9325	MULT VEL					
	2314	N	101-200	KI		
	2312	E	101-200	SN	8	
9328						
	1425	E	101-200	SN	8	
9347						
	2140	N	101-200	KI		
9352	MULT VEL					
	2249	E	301-400	V	5	
	2130	E	301-400	A	4	
9353						
	1371	N	21-CM	S	3	
9354						
	3701	E	401-500	MW		
9358						
	2300	A	401-500	L		

143

UGC	NGC	R.A. (1950) DEC	m pg	Hubble Type	Char.	Cluster	Group
9361	I1029	14 30.7 +50 07	13.7	SB			
9363	N5668	14 30.9 +04 40	12.7	SC			
9366	N5676	14 31.0 +49 40	11.7	SC			
9388	N5682	14 33.0 +48 53	15.1	SB	DISR		9399
9391		14 33.2 +59 34	15.5	••• SC-I		329	
9395	N5687	14 33.3 +54 42	13.3	S0			
9399	N5689	14 33.7 +48 57	12.7	E SA S0		BR IN GRP	9399
9405		14 34.0 +57 28	17.0	IRR			
9416	N5690	14 35.2 +02 28	13.1	SC		333	
9427	N5692	14 35.8 +03 37	13.3	PEC		332	
9436	N5701	14 36.7 +05 35	12.9	S0 SA			
9451	N5713	14 37.6 -00 05	11.7	SC SB		BR IN GRP	9451
9486	N5739	14 40.6 +42 03	13.7	S0-A S•••			
9493	N5740	14 41.9 +01 54	13.2	SB		333	
9499	N5746	14 42.5 +02 10	12.3	SB		333	
9500		14 42.8 +08 05	18.0	SC-I			
9509		14 43.5 +08 43	15.4	DBLE		330	9509
9525	N5759	14 44.9 +13 40	14.9	DBLE	DISR	336	
9553	I1065	14 48.2 +63 29	15.0	S0 SA			
9558		14 48.8 +17 24	15.0	SC		339	
9560		14 48.9 +35 47	14.5	PEC			
9561		14 49.0 +09 32	14.9	•••	DISR		
9562		14 49.2 +35 45	14.2	PEC	DISR		

UGC	V_r (km s^{-1})	Spectrum	Disp. (A/mm)	Ref.	Notes
9361					
	2377	N	101-200	K I	
9363	MULT VEL				
	1550	E	101-200	FR 8	
	1780	E	401-500	MW	
	1665	E	401-500	L	
	1581	N	21-CM	DU	
	1577	N	21-CM	R 1	
9366	MULT VEL				
	2244	A	401-500	L	
	2196	N	21-CM	BG 3	
	2225	N	21-CM	BA 1	
	2157	N	101-200	K I	
9388	MULT VEL				
	2283	E	301-400	V 5	
	2220	N	201-300	A 7	
9391					
	1920	N	21-CM	FT	
9395					
	2119	A	401-500	MW	
9399	MULT VEL				
	2205	A	401-500	MW	
	2028	N	101-200	K I	
9405	MULT VEL				MAG NOT ZW
	222	N	21-CM	FT	
	215	N	21-CM	AL 3	
9416					
	2190	A	301-400	V 8	
9427					
	1800	E	201-300	AD10	
9436					
	1556	A	101-200	SN 8	
9451	MULT VEL				
	1870	E	401-500	MW	
	1965	E	401-500	L	
	1865	E	101-200	FR 8	
	1862	N	21-CM	BG 2	
	1875	N	21-CM	FR 8	
	1801	N	21-CM	R 3	
9486					
	5578	E	101-200	SN 8	
9493					
	1772	E	101-200	SN 8	
9499	MULT VEL				
	1789	E	401-500	MW	
	1964	A	301-400	LB	
	1882	A	401-500	L	
9500					MAG NOT ZW
	1691	N	21-CM	FT	
9509					
	10470	E	301-400	BB28	
9525					
	8400	E	201-300	AD10	
9553	MULT VEL				3C305
	12343	E	301-400	SN 8	
	12600	E	201-300	DE	
9558					
	13581	E	101-200	FR11	
9560	MULT VEL				
	1015	E	301-400	SA 2	
	1282	E	1-100	CK	
	1209	N	21-CM	LA	
9561					MAG SHARED
	8776	E	301-400	KR	
9562	MULT VEL				
	1200	E	301-400	SA 2	
	1238	N	21-CM	LA	

UGC	NGC	R.A. (1950) DEC	m pg	Hubble Type	Char.	Cluster	Group
9576	N5774	14 51.1 +03 47	13.9	SC			
9579	N5775	14 51.5 +03 45	13.0	SC			
9587	I4516	14 52.0 +16 34	14.9	E		339	
9593	I1075	14 52.5 +18 18	14.9	SB			
9595	I1076	14 52.6 +18 14	13.9	•••			
9599	N5787	14 53.4 +42 42	14.1	•••			
9602		14 53.5 +12 04	14.9	SO			
9631	N5792	14 55.8 -00 55	13.5	SB			
9642	N5820	14 57.1 +54 05	13.0	E-SO SO	DISR	345 BR IN GRP	9642
9645	N5806	14 57.4 +02 05	12.9	SB		352	9706
9649	N5832	14 57.6 +71 53	13.3	SC			
9655	N5813	14 58.6 +01 53	12.5	E		352	9706
9663		14 59.6 +52 48	15.2	IRR		344	
9673	N5829	15 00.4 +23 31	14.6	SC			
9678	N5831	15 01.6 +01 24	13.1	E		352	
9692	N5838	15 02.8 +02 16	12.1	SO		352	
9700	N5845	15 03.5 +01 48	13.8	E		352	9706
9706	N5846	15 04.0 +01 46	11.9	E		352 BR IN GRP	9706
9714	N5851	15 04.5 +13 04	14.9	•••	DISR		9686
9715	N5850	15 04.7 +01 44	13.6	SB		352	9706
9717	N5860	15 04.7 +42 50	14.2	DBLE	MARK		
9723	N5866	15 05.1 +55 57	11.1	SO		344	
9724	N5857	15 05.2 +19 47	13.6	SA SB		345	
9726	N5854	15 05.3 +02 45	13.1	SO SA		352	
9728	N5859	15 05.3 +19 46	13.1	SB		345	
9736	N5874	15 06.4 +54 56	14.1	SC		344	

UGC	V_r (km s^{-1})	Spectrum	Disp. (Å/mm)	Ref.	Notes
9576	MULT VEL				
	1542	E	301-400	P	
	1680	N	21-CM	GU	
9579					
	1582	E	301-400	P	
9587					3C306
	13680	A	101-200	T	
9593					
	6259	E	301-400	KR	
9595	MULT VEL				
	6339	E	301-400	KR	
	6190	E	NO INFO	DS 7	
9599					
	5524	A	301-400	SA 2	
9602					MC 1453+12
	9600	A	201-300	BS	
9631					
	1985	E	301-400	V 5	
9642					
	3269	A	401-500	MW	
9645					
	1301	A	401-500	MW	
9649					
	600	E	NO INFO	BB44	
9655					
	1882	A	401-500	MW	
9663					
	2422	N	21-CM	FT	
9673					MAG SHARED
	5689	E	101-200	FR11	
9678					
	1684	A	401-500	MW	
9692					
	1427	A	401-500	MW	
9700					
	1785	A	301-400	V 7	
9706	MULT VEL				MAG SHARED
	1768	E	401-500	MW	
	1774	E	401-500	L	
	1795	N	101-200	TU	
	1826	N	21-CM	HT	
9714					
	16389	N	101-200	TU	
9715	MULT VEL				
	2319	E	401-500	MW	
	2476	E	401-500	L	
9717	MULT VEL				
	5400	E	201-300	AD 6	
	5460	E	301-400	KR	COMP VEL A
	5530	E	301-400	KR	COMP VEL B
9723	MULT VEL				
	740	E	401-500	MW	
	850	E	401-500	L	
	650	N	NO INFO	SB	
	690	A	101-200	SI	
	692	A	1-100	V 6	
9724	MULT VEL				
	4616	A	401-500	MW	
	4721	A	401-500	L	
	4602	N	101-200	TU	
9726					
	1626	A	401-500	MW	
9728	MULT VEL				
	4664	A	401-500	MW	
	4605	N	101-200	TU	
9736					
	3315	N	21-CM	FR11	

UGC	NGC	R.A. (1950) DEC				m_{pg}	Hubble Type	Char:	Cluster	Group
9740	N5864	15	07.0	+03	15	12.9	S0		352	
9753	N5879	15	08.4	+57	12	11.9	SB			
9789	N5899	15	13.2	+42	14	12.6	SC SB		351	
9797	N5905	15	14.0	+55	42	13.6	SB			
9799		15	14.3	+07	12	14.8	E		BR IN GRP	9799
9801	N5907	15	14.6	+56	30	11.4	SC SB			
9805	N5908	15	15.3	+55	35	13.5	SB SA-B			
9822	N5920	15	19.4	+07	53	15.5	S0 E-S0			
9823	N5923	15	19.4	+41	54	14.7	SB-C SC		351	
9824	N5921	15	19.5	+05	15	12.7	SB			
9837		15	22.6	+58	14	14.6	SC			
9846		15	23.7	+16	30	14.7	SB-C SC			
9851	N5929	15	24.3	+41	51	13.0	E-S0 E		362	
9852	N5930	15	24.3	+41	51	13.0	SA E		362	
9866	N5949	15	27.4	+64	55	12.7	SC			
9867	N5936	15	27.7	+13	10	13.0	SB SC			
9893		15	31.3	+46	37	15.2	PEC DBLE			
9903	N5953	15	32.2	+15	21	13.3	S0 E	DISR		
9904	N5954	15	32.2	+15	21	13.7	SC	DISR		
9918	N5961	15	33.2	+31	01	14.0	...		364	
9922		15	34.0	+38	50	14.4	DBLE			
9926	N5962	15	34.2	+16	46	12.2	SC			
9928	I4562	15	34.3	+43	39	13.8	E		362	9940
9932	I1143	15	35.0	+82	38	14.7	E		388	
9933	I4566	15	35.0	+43	43	14.3	SB SC		362	9940
9936		15	35.1	+44	24	15.5	SC-I			

UGC	V_r (km s^{-1})	Spectrum	Disp. (A/mm)	Ref.	Notes
9740					
	1618	A	301–400	V 5	
9753	MULT VEL				
	876	E	401–500	MW	
	790	N	21–CM	BA 1	
9789					
	2549	E	401–500	MW	
9797	MULT VEL				
	3244	E	201–300	BC	
	3356	E	201–300	SN 8	
9799	MULT VEL				3C317
	10500	E	101–200	PT	
	10354	N	101–200	TU	
	10143	A	201–300	MS	
9801	MULT VEL				
	553	A	401–500	MW	
	522	A	401–500	L	
	660	N	21–CM	GU	
9805					
	3403	A	201–300	SN 8	
9822	MULT VEL				3C318.1
	13146	A	101–200	SN 8	
	13800	A	1–100	SL	
					DISCREPANT VELOCITIES?
9823					
	5571	E	1–100	FR11	
9824	MULT VEL				
	1389	A	401–500	MW	
	1480	N	21–CM	GU	
9837					
	2674	N	101–200	FR11	
9846					
	7007	E	101–200	FR11	
9851	MULT VEL				MAG SHARED
	2522	E	301–400	P	
	2436	N	101–200	TU	
9852	MULT VEL				MAG SHARED
	2698	E	301–400	P	
	2714	N	101–200	TU	
9866	MULT VEL				
	380	E	401–500	L	
	571	E	101–200	SN 8	
9867					
	4108	E	101–200	SN 8	
9893					
	645	E	301–400	SA 2	
9903	MULT VEL				
	2164	E	301–400	P	
	2048	N	101–200	TU	
9904	MULT VEL				
	2123	E	301–400	P	
	1956	N	101–200	TU	
9918					
	1800	E	201–300	AD 8	
9922					
	5493	E	301–400	SA 2	
9926					
	1993	A	401–500	MW	
9928					
	5860	A	NO INFO	Z 3	
9932					
	6619	A	201–300	GY	
9933					
	5588	E	1–100	FR11	
9936					
	2628	N	21–CM	FT	

UGC	NGC	R.A. (1950) DEC	m_{pg}	Hubble Type	Char.	Cluster	Group
9943	N5970	15 36.1 +12 21	12.2	SC			
9948	N5981	15 36.8 +59 33	14.2	SB-C			9969
9952	N5974	15 37.0 +31 55	14.3	...		364	
9961	N5982	15 37.6 +59 31	12.4	E			9969
9969	N5985	15 38.6 +59 30	12.0	SB			9969 BR IN GRP
9979		15 39.8 +00 37	15.7	IRR			
10003	N5992	15 42.6 +41 15	14.2	S...	MARK	362	
10007	N5993	15 42.7 +41 17	13.9	SB		362	
10033	N5994	15 44.7 +18 02	13.2	DBLE SB	DISR		
10054		15 48.0 +81 59	14.9	... SC-I		388	
10061		15 48.8 +16 27	16.5	IRR			
10075	N6015	15 50.6 +62 27	11.7	SC			
10079	N6007	15 51.0 +12 06	14.1	SC			
10083	N6012	15 51.9 +14 44	13.1	SA			
10086		15 52.3 +16 45	14.5	...	DISR		
10099		15 54.9 +42 01	14.3	COMP		378	
10101	N6018	15 55.2 +16 00	14.6	SO		382	
10102	N6021	15 55.2 +16 05	14.1	E		382	
10104		15 55.4 +30 12	14.9	SC			
10106	N6023	15 55.5 +16 27	14.7	E		382	10106
10116	N6027	15 57.0 +20 55	13.4	MULT SA COMP COMP SC			BR IN GRP 10116
10120		15 57.3 +35 10	14.9	SB	MARK		

UGC	V_r (km s^{-1})	Spectrum	Disp. (A/mm)	Ref.	Notes
9943	MULT VEL				
	2034	A	401-500	MW	
	2127	A	401-500	L	
9948					
	1764	A	301-400	V 8	
9952					
	1800	E	201-300	AD10	
9961	MULT VEL				
	2864	A	401-500	MW	
	2981	A	401-500	L	
9969					
	2467	E	401-500	MW	
9979					
	1978	N	21-CM	FT	
10003	MULT VEL				
	9300	E	201-300	AD 7	
	10200	E	301-400	DS 4	
	9414	E	301-400	KR	DISCREPANT VELOCITIES?
10007	MULT VEL				
	9591	E	301-400	KR	
	9396	N	101-200	TU	
10033	MULT VEL				
	3309	E	301-400	KR	COMP VEL A
	3064	E	301-400	KR	COMP VEL B
10054					
	1499	N	21-CM	FT	
10061					MAG NOT ZW
	2087	N	21-CM	FT	
10075	MULT VEL				
	822	E	1-100	CA	
	646	E	401-500	MW	
	732	E	401-500	L	
	850	N	21-CM	GU	
	835	N	21-CM	S 3	
	852	N	21-CM	R 3	
	675	N	NO INFO	HJ	
10079					
	10487	E	101-200	FR11	
10083					
	1990	N	NO INFO	HJ	
10086					
	2400	E	201-300	AD10	
10099	MULT VEL				
	10334	E	301-400	SA 2	
	10592	E	1-100	CK	
	10200	E	201-300	AD10	
10101					
	5121	E	201-300	BZ	
10102					
	4486	A	201-300	BZ	
10104					
	9923	N	101-200	FR11	
10106					
	11140	A	201-300	BZ	
10116	MULT VEL				
	4031	A	401-500	MW	COMP VEL A
	4036	A	401-500	L	COMP VEL A
	4285	A	201-300	CR 3	COMP VEL A
	4187	A	201-300	CR 3	COMP VEL B
	4438	A	201-300	CR 3	COMP VEL C
	4415	A	401-500	MW	COMP VEL D
	4468	A	401-500	L	COMP VEL D
	4392	A	201-300	CR 3	COMP VEL D
10120					
	9900	E	301-400	DS 4	

UGC	NGC	R.A. (1950) DEC	m_{pg}	Hubble Type	Char.	Cluster	Group
10126	N6068	15 58.0 +79 08	13.3	S... SB-C	DISR		
10128		15 58.4 +30 31	15.2	DBLE	MARK		
10135	N6028	15 59.2 +19 29	14.8	SO		382	
10143		16 00.0 +16 06	14.9	E		382 BR IN GRP	10143
10144		16 00.0 +16 29	14.6	E		382	
10148	N6032	16 00.8 +21 06	15.0	S... SB		382	
10154	N6035	16 01.2 +21 02	14.7	SC		382	
10165	N6040	16 02.1 +17 53	14.6	DBLE		382	
10170	N6041	16 02.3 +17 51	14.9	E SO		382	
10177	N6045	16 02.8 +17 54	14.8	SB		382	
10178	N6051	16 02.8 +24 04	14.9	E		382	
10180	I1173	16 02.9 +17 33	15.6	S...		382	
10182	N6052	16 03.0 +20 41	14.1	DBLE SB S... S... SC-I	MARK	382	
10186	N6050	16 03.1 +17 54	14.9	DBLE SB-C		382	
10187		16 03.2 +16 35	15.4	DBLE E		382	
10188	I1178	16 03.2 +17 44	15.0	DBLE SO-A		382	
10189	I1181	16 03.2 +17 44	15.0	DBLE SO-A		382	
10191	N6055	16 03.2 +18 17	15.4	SO SO-A		382	
10192	I1182	16 03.3 +17 56	15.2	PEC	SEYF	382	
10196	N6060	16 03.7 +21 38	14.3	SC		382	
10199	N6061	16 04.0 +18 23	15.0	SO		382	
10200		16 04.0 +41 29	13.6	... SO-A			

UGC	V_r (km s^{-1})	Spectrum	Disp. (A/mm)	Ref.		Notes
10126	MULT VEL					
	3925	E	301-400	P		
	4200	E	201-300	AD 8		
10128						
	9510	E	NO INFO	DS 6		
10135	MULT VEL					
	4522	A	101-200	CM		
	4480	E	NO INFO	Z 3		
10143	MULT VEL					
	10483	A	201-300	BZ		
	10475	N	NO INFO	BN		
	10440	A	101-200	PT		
10144						
	11449	A	201-300	BZ		
10148						
	4284	E	101-200	CM		
10154						
	2241	A	101-200	CM		
10165	MULT VEL					1602+17W1
	12386	E	201-300	CR 3	COMP VEL	A
	12618	E	201-300	CR 3	COMP VEL	B
	12283	E	201-300	UL16		
10170	MULT VEL					
	10469	A	401-500	MW		
	10530	A	301-400	BB 4		
10177	MULT VEL					4C17.66
	10440	A	101-200	PT		
	9935	A	401-500	MW		
	9943	A	301-400	BB 4		
10178	MULT VEL					4C24.36
	9540	A	101-200	SA 4		
	9690	A	NO INFO	SP		
10180						
	10840	A	301-400	BB 4		
10182	MULT VEL					
	4671	E	301-400	V 5		
	4800	E	201-300	AD 3		
	4500	N	101-200	TU	COMP VEL	A
	4541	N	101-200	TU	COMP VEL	B
	4680	E	201-300	HS		
	4731	E	101-200	DP 2		
	4770	E	NO INFO	WD 2		
	4744	E	1-100	V 6		
	4730	E	201-300	SN 8		
10186						
	11057	A	301-400	BB 4		
10187						
	13080	A	101-200	PT		
10188					MAG SHARED	
	10324	A	301-400	BB 4		
10189					MAG SHARED	
	10690	A	301-400	BB 4		
10191	MULT VEL					
	11390	E	101-200	MK		
	11314	E	301-400	BB 4		
10192	MULT VEL					
	10245	E	301-400	BB 4		
	10200	E	201-300	AD 3		
	10180	E	NO INFO	SK 1		
	10230	E	NO INFO	WD 2		
10196						
	4555	A	101-200	CM		
10199						
	11169	A	301-400	BB 4		
10200	MULT VEL					
	1947	N	101-200	TU		
	1980	E	301-400	KR		

153

UGC	NGC	R.A. (1950) DEC	m_{pg}	Hubble Type	Char.	Cluster	Group
10201		16 04.1 +15 49	14.3	MULT E E		382	
10230	N6070	16 07.4 +00 50	13.0	SC		385	
10267	N6090	16 10.4 +52 34	14.0	DBLE IRR IRR	MARK		
10270	N6086	16 10.5 +29 38	14.8	E		386	
10310		16 14.8 +47 10	14.9	IRR			
10311	N6107	16 15.4 +35 01	14.7	E		393	
10316	N6109	16 15.7 +35 07	14.9	COMP		393	
10325		16 16.0 +46 13	14.5	DBLE			
10328	N6106	16 16.4 +07 32	13.4	SC			
10336	N6116	16 17.0 +35 17	15.3	SB			
10339		16 17.6 +02 08	14.3	...			
10343	N6120	16 18.0 +37 54	14.3	PEC		391	
10350	N6118	16 19.3 −02 10	13.2	SC			
10357		16 20.3 +40 34	15.1	...		395	
10364	N6137	16 21.3 +38 02	14.1	E		391	
10385	N6155	16 24.7 +48 28	13.0	S...		389	
10400	N6160	16 26.0 +41 02	14.8	E		395	
10407		16 26.8 +41 20	14.3	TRPL			
10409	N6166	16 26.9 +39 40	13.9	E		395 BR IN CLUS	
10421	N6173	16 28.1 +40 55	14.0	E		395	
10430		16 28.9 +41 36	15.4	SB		395	
10439	N6181	16 30.1 +19 56	12.7	SC			
10442	N6189	16 30.8 +59 45	13.3	SC		403	
10443	N6190	16 31.2 +58 33	13.2	SC		403	
10461	I1222	16 33.6 +46 19	14.6	SB			
10465		16 34.2 +01 45	14.7	SB-C			

154

UGC	V_r (km s^{-1})	Spectrum	Disp. (A/mm)	Ref.	Notes
10201	MULT VEL				
	12071	N	101–200	TU	COMP VEL A
	12344	N	101–200	TU	COMP VEL B
10230	MULT VEL				
	2120	E	401–500	L	
	2091	A	401–500	MW	
	2010	N	21–CM	S 3	
10267	MULT VEL				
	8733	E	301–400	SA 2	
	8700	E	201–300	AD 7	
	8690	N	101–200	TU	COMP VEL A
	8883	E	301–400	KR	COMP VEL A
	8612	N	101–200	TU	COMP VEL B
	8927	E	301–400	KR	COMP VEL B
	8780	E	NO INFO	DS 6	
10270					
	9390	A	101–200	PT	
10310					
	715	N	21–CM	FT	
10311					
	9182	A	201–300	UL18	
10316					
	8850	A	201–300	UL18	
10325	MULT VEL				
	5885	E	301–400	KR	COMP VEL A
	5865	E	301–400	KR	COMP VEL B
10328					
	1470	E	301–400	V 5	
10336					
	8800	A	201–300	UL18	
10339					
	5100	E	201–300	AD10	
10343	MULT VEL				
	9367	E	201–300	CR 2	
	9332	E	301–400	SA 2	
	9290	E	NO INFO	Z 3	
10350					
	1571	N	21–CM	S 3	
10357					
	9308	A	201–300	CR 2	
10364					
	9310	A	201–300	UL18	
10385					
	2424	E	101–200	PR	
10400	MULT VEL				
	9450	A	101–200	PT	
	9344	A	201–300	CR 2	
10407	MULT VEL				
	8502	E	201–300	CR 2	
	8700	E	201–300	AD 8	
10409					MAG SHARED 3C338
	8882	E	301–400	MK	
10421					
	8727	A	201–300	CR 2	
10430					
	9072	A	201–300	CR 2	
10439	MULT VEL				
	2158	A	701–1000	MW	
	2350	E	1–100	BB39	
10442					
	5533	E	1–100	PR	
10443					
	3416	E	1–100	PR	
10461					
	9298	N	101–200	FR11	
10465					
	7358	E	101–200	FR11	

UGC	NGC	R.A. (1950) DEC	m pg	Hubble Type	Char.	Cluster	Group
10469	N6195	16 34.8 +39 08	14.7	SB		395	
10470	N6217	16 34.9 +78 18	12.1	SB SC		411	
10491		16 36.7 +42 02	15.5	DBLE E E			
10510		16 39.6 +58 13	14.9	SC		403	
10521	N6207	16 41.3 +36 55	11.9	S... SC			
10577	N6239	16 48.5 +42 50	12.9	PEC SB			
10592	N6240	16 50.5 +02 29	14.7	PEC			
10599		16 52.2 +39 50	13.7	COMP		420	
10608		16 53.3 +53 13	16.5	... SC-I			
10633	I1236	16 56.3 +20 07	14.6	SC			
10635		16 56.3 +38 17	13.5	COMP			
10665	N6290	17 00.1 +59 04	14.3	SA		403 BR IN	10665 GRP
10724	N6306	17 07.0 +60 47	14.3	S...		403	
10727	N6307	17 07.1 +60 48	14.0	SO-A E		403	
10752	N6314	17 10.5 +23 20	14.3	SA		BR IN	10752 GRP
10762	N6340	17 11.3 +72 21	11.9	SA SO-A		430 BR IN	10762 GRP
10804	N6359	17 17.4 +61 50	13.6	SO E			
10805		17 17.6 +14 27	15.3	SC-I			
10814		17 18.0 +49 05	14.9	S...			
10855	N6376	17 24.6 +58 52	14.1	DBLE	DISR	429	
10861	N6372	17 25.5 +26 30	14.1	S... SC	DISR		
10886	N6379	17 28.4 +16 19	14.6	SC			10875
10891	N6384	17 30.0 +07 06	13.2	SB			
10897	N6412	17 31.2 +75 45	12.4	SC			
10937	I1267	17 38.0 +59 25	14.5	SB SC		429	

UGC	V_r (km s^{-1})	Spectrum	Disp. (A/mm)	Ref.	Notes
10469					
	9000	E	1-100	FR11	
10470	MULT VEL				
	1386	E	201-300	MW	
	1382	E	401-500	L	
	1363	N	21-CM	DL 2	
	1325	N	21-CM	BA 1	
	1356	N	21-CM	R 3	
	1355	N	21-CM	PS	
10491	MULT VEL				
	8572	E	301-400	A 1	COMP VEL A
	8118	E	301-400	A 1	COMP VEL B
					DISCREPANT VELOCITIES?
10510					
	5405	E	101-200	FR11	
10521	MULT VEL				
	835	E	1-100	CA	
	869	E	401-500	MW	
	852	N	21-CM	BG 3	
10577	MULT VEL				
	964	E	401-500	L	
	931	N	21-CM	BG 3	
10592					PK1650+024
	7500	E	101-200	WH	
10599	MULT VEL				4C39.49
	9936	A	1-100	UL14	
	10050	A	201-300	WL	
10608					MAG NOT ZW
	1094	N	21-CM	FT	
10633					
	6073	E	101-200	FR11	
10635					
	10075	A	NO INFO	Z 3	
10665					
	12300	E	201-300	AD 8	
10724	MULT VEL				
	3064	E	301-400	V 5	
	2820	E	101-200	KK	
10727	MULT VEL				
	3283	E	301-400	V 5	
	2820	A	101-200	KK	
					DISCREPANT VELOCITIES?
10752					
	6748	E	401-500	L	
10762	MULT VEL				
	2109	A	701-1000	MW	
	1902	N	21-CM	BG 3	
10804					
	2948	A	201-300	MW	
10805					
	1557	N	21-CM	FT	
10814					
	7182	E	101-200	ZH 2	
10855					
	8721	E	301-400	KR	
10861					
	4739	E	101-200	FR11	
10886					
	5964	N	101-200	FR11	
10891	MULT VEL				
	1781	A	301-400	LB	
	1717	A	401-500	L	
	1784	A	401-500	MW	
10897					
	1508	E	401-500	L	
10937					
	9311	E	101-200	FR11	

UGC	NGC	R.A. (1950) DEC	m_{pg}	Hubble Type	Char.	Cluster	Group
10998	N6478	17 47.5 +51 12	14.1	SC			
11000		17 47.7 +36 09	14.0	...			
11009	N6482	17 49.7 +23 05	12.8	E			
11012	N6503	17 49.9 +70 10	10.9	SC			
11013	I1269	17 50.0 +21 35	14.5	SB-C SC	DISR		
11041		17 53.0 +34 47	13.9	SA-B			
11048	N6500	17 53.8 +18 21	13.4	SA	DISR	436	
11058		17 55.0 +32 38	14.4	SB SC			
11117	N6560	18 03.9 +46 52	14.2	SB SC			
11127		18 06.8 +28 02	14.8	DBLE	DISR	438	
11130		18 07.2 +69 49	14.4	COMP N		443	
11137	N6570	18 08.8 +14 05	13.2	SC-I			
11144	N6574	18 09.6 +14 57	12.5	S... SB			
11175	N6621	18 13.2 +68 20	13.6	DBLE			
11212	N6627	18 20.4 +15 40	14.5	SB			
11218	N6643	18 21.2 +74 32	11.8	SC SB			
11221	N6636	18 22.0 +66 36	14.2	DBLE	DISR		
11238	N6654	18 25.2 +73 09	12.7	S0 SA			
11239	N6635	18 25.3 +14 44	14.5	S0			
11252		18 27.4 +48 13	14.9	SC		453	
11274	N6658	18 31.8 +22 51	15.0	S0-A S0	DISR		
11282	N6661	18 32.5 +22 52	14.1	S0			
11308	N6674	18 36.5 +25 20	13.7	SB			
11318	N6691	18 38.3 +55 35	14.1	SB SC			
11354	N6702	18 45.6 +45 40	13.8	E		453	
11356	N6703	18 45.9 +45 30	12.4	S0		453	

UGC	V_r (km s^{-1})	Spectrum	Disp. (A/mm)	Ref.	Notes
10998					
	6857	A	401-500	MW	
11000					
	300	E	201-300	AD 8	
11009					
	3922	A	401-500	MW	
11012	MULT VEL				
	13	E	301-400	L B	
	33	E	401-500	L	
	60	E	301-400	BB30	
	60	N	21-CM	GU	
	62	N	21-CM	S 3	
	80	N	21-CM	R 3	
11013					
	6072	E	101-200	FR11	
11041					
	5100	E	201-300	AD 8	
11048					
	2950	E	101-200	KR	
11058					
	4757	N	101-200	FR11	
11117					
	7034	E	101-200	FR11	
11127					
	6900	E	201-300	AD 8	
11130					3C371
	14970	E	301-400	SN 8	
11137					
	1975	A	301-400	V 8	
11144	MULT VEL				
	2355	E	401-500	MW	
	2387	E	401-500	L	
	2270	E	101-200	UL 4	
11175	MULT VEL				
	6230	E	301-400	BB28	
	6194	E	301-400	KR	COMP VEL A
	5941	E	301-400	KR	COMP VEL B
11212					
	5206	E	401-500	MW	
11218	MULT VEL				
	1494	E	401-500	MW	
	1682	E	401-500	L	
	1507	N	21-CM	BG 3	
	1477	N	21-CM	R 3	
11221	MULT VEL				
	4193	E	301-400	KR	COMP VEL A
	4366	E	301-400	KR	COMP VEL B
11238					
	1924	A	401-500	L	
11239					
	5071	A	401-500	L	
11252					
	4930	E	101-200	KZ	
11274					
	4270	E	401-500	MW	
11282	MULT VEL				
	4370	A	401-500	MW	
	4193	A	401-500	L	
11308					
	3502	A	401-500	MW	
11318					
	5858	E	101-200	FR11	
11354	MULT VEL				
	4749	A	401-500	MW	
	4706	A	401-500	L	
11356	MULT VEL				
	2394	E	401-500	L	
	2316	A	401-500	MW	

UGC	NGC	R.A. (1950) DEC	m pg	Hubble Type	Char.	Cluster	Group
11361	N6711	18 47.6 +47 36	14.1	SB-C SC		453	
11364	N6710	18 48.6 +26 46	14.6	SA			
11406		19 06.8 +43 00	14.7	S••• SC		453	
11407	N6764	19 07.0 +50 50	13.2	SB	SEYF	453	
11414	N6786	19 11.9 +73 18	13.7	S•••	DISR		
11438	I1301	19 25.3 +49 39	14.7	SC		453	
11453		19 29.9 +54 00	14.4	SB SC	DISR	453	
11465		19 40.4 +50 32	14.4	••• E		453	
11470	N6824	19 42.6 +55 59	13.1	SA-B SB		453	
11492		19 51.6 +57 20	14.2	SB-C SC		453	
11503		19 57.2 +49 54	14.5	••• E		453	
11524		20 09.6 +05 37	14.9	SC	DISR		
11546	I1317	20 20.7 +00 30	14.5	COMP SO			
11555		20 22.8 +05 06	14.7	SB-C SC	DISR		
11562		20 24.2 +02 32	14.7	SB SC			11560
11570	N6921	20 26.4 +25 33	15.0	SB SA			
11585		20 29.7 -02 24	14.8	SB		458	
11589	N6928	20 30.5 +09 45	13.7	SA-B SA		BR IN GRP	11589
11590	N6930	20 30.6 +09 41	14.3	SB			11589
11597	N6946	20 33.8 +59 59	10.5	SC			
11604	N6951	20 36.6 +65 55	12.3	SB-C SB			
11618	N6954	20 41.6 +03 02	14.2	S•••			
11628	N6962	20 44.8 +00 08	13.5	SB SA		BR IN GRP	11628
11629	N6964	20 44.9 +00 07	14.2	SB E			11628
11633	N6969	20 46.0 +07 33	15.0	SA			
11640	N6972	20 47.6 +09 43	14.3	SO-A E		BR IN GRP	11640
11643		20 49.2 +18 46	14.8	SB	DISR		
11657		20 57.2 -02 04	14.4	DBLE	DISR	458	
11658		20 57.2 -02 04	14.4	DBLE	DISR	458	

UGC	V_r (km s^{-1})	Spectrum	Disp. (A/mm)	Ref.	Notes
11361					
	4639	E	101-200	FR11	
11364					
	4556	A	401-500	MW	
11406					
	4562	E	101-200	FR11	
11407					
	2405	E	1-100	FR 9	
11414					
	7997	E	301-400	KR	
11438					
	3969	E	101-200	FR11	
11453					MAG SHARED
	3922	E	101-200	FR11	
11465	MULT VEL				3C402
	7120	A	301-400	BS	
	7872	A	101-200	SN 8	
					DISCREPANT VELOCITIES?
11470					
	3386	E	201-300	MW	
11492					
	3602	E	101-200	FR11	
11503					
	7336	A	101-200	SN 8	
11524					
	5232	E	1-100	FR11	
11546					
	3975	A	401-500	L	
11555					
	4760	E	101-200	FR11	
11562					
	5289	E	101-200	FR11	
11570					MAG NOT ZW
	4317	E	401-500	MW	
11585					
	5943	E	1-100	FR11	
11589					
	4754	A	401-500	MW	
11590					
	4182	E	401-500	MW	
11597	MULT VEL				
	38	E	401-500	MW	
	-70	E	401-500	L	
	50	N	21-CM	DD	
	40	N	21-CM	RS	
	45	N	21-CM	RR	
	50	N	21-CM	GR	
11604	MULT VEL				
	1364	A	401-500	L	
	1396	N	21-CM	BG 3	
	1380	N	21-CM	BA 1	
11618					
	4011	A	401-500	MW	
11628					
	4183	E	401-500	MW	
11629					
	3832	A	401-500	MW	
11633					
	4000	A	101-200	KD	
11640					
	4442	A	301-400	V 5	
11643					
	8714	E	301-400	KR	
11657					MAG SHARED
	5883	E	301-400	KR	
11658					MAG SHARED
	5898	E	301-400	KR	

UGC	NGC	R.A. (1950) DEC	m_{pg}	Hubble Type	Char.	Cluster	Group
11668		21 00.4 +36 30	14.0	MULT COMP			
11670	N7013	21 01.4 +29 42	12.9	SA SO-A			
11680		21 05.2 +03 40	14.5	DBLE SC COMP			
11695		21 09.6 −01 40	14.7	SB SC	DISR		
11718	N7052	21 16.3 +26 14	14.0	E			
11756	N7080	21 27.8 +26 30	14.1	SB			
11810	I1401	21 44.4 +01 28	14.7	SB SC			
11815	N7137	21 45.9 +21 56	13.3	S... SC			
11830		21 48.8 +25 37	15.0	SB			
11843	N7156	21 52.0 +02 42	13.5	SC SC-I		464	
11865		21 56.2 +11 48	14.3	...	MARK		
11871		21 58.2 +10 19	14.7	DBLE	MARK		
11872	N7177	21 58.3 +17 30	12.2	SB			
11880	I1420	22 00.2 +19 30	14.5	S... PEC PEC			
11914	N7217	22 05.6 +31 07	11.0	SB			
11919		22 06.1 +40 56	14.8	SB-C SC		467	11931
11946		22 09.6 +46 04	14.7	SC			
11958	N7236	22 12.3 +13 35	14.3	TRPL		469	
11969	N7242	22 13.4 +37 03	14.6	E		468 BR IN GRP	11969
11991		22 18.3 +47 27	14.4	SC			
12011		22 20.8 +30 40	14.0	DBLE COMP COMP	DISR		
12026	N7274	22 22.0 +35 53	14.2	E		471 BR IN GRP	12026
12035	N7280	22 24.1 +15 54	13.6	SO-A SO			
12048	N7292	22 26.1 +30 03	13.1	IRR			
12064		22 29.1 +39 06	14.6	E-SO E		471 BR IN GRP	12064
12066		22 29.4 +19 25	14.6	DBLE	MARK	473	

UGC	V_r (km s^{-1})	Spectrum	Disp. (A/mm)	Ref.	Notes
1668					
	3000	A	301-400	C A	
1670	MULT VEL				
	570	E	301-400	V 5	
	852	N	21-CM	BG 3	
1680	MULT VEL				
	7854	E	301-400	KR	COMP VEL A
	7982	E	301-400	KR	COMP VEL B
	7680	E	301-400	SA 2	
1695					
	9684	E	101-200	FR11	
1718					
	4920	A	NO INFO	UL17	
1756					
	4806	E	101-200	FR11	
1810					
	4648	E	101-200	FR11	
1815	MULT VEL				
	1505	A	401-500	L	
	1654	N	21-CM	BI 1	
1830					
	5696	E	101-200	FR11	
1843					
	3966	A	301-400	V 5	
1865					
	9750	E	301-400	DS 4	
1871					
	8400	E	301-400	DS 4	
1872	MULT VEL				
	1105	E	401-500	MW	
	1225	N	21-CM	BG 3	
1880	MULT VEL				
	1413	E	301-400	KR	COMP VEL A
	1476	E	301-400	KR	COMP VEL B
11914	MULT VEL				
	950	E	101-200	FR 8	
	911	E	201-300	MW	
11919					
	5313	E	101-200	FR11	
11946					
	5472	E	101-200	FR11	
11958	MULT VEL				3C442
	7857	E	101-200	G	COMP VEL 6
	7867	A	201-300	J	COMP VEL 6
	7852	E	101-200	G	COMP VEL 7
	7866	A	201-300	J	COMP VEL 7
11969					MAG SHARED
	5684	A	401-500	MW	
11991					
	5567	E	101-200	FR11	
12011	MULT VEL				
	6702	E	201-300	CR 2	COMP VEL A
	6646	E	301-400	KR	COMP VEL A
	6717	E	201-300	CR 2	COMP VEL B
	6552	E	301-400	KR	COMP VEL B
12026					
	5820	A	101-200	PT	
12035	MULT VEL				
	1812	A	301-400	V 5	
	1858	A	301-400	V 8	
	1857	N	21-CM	BI 1	
12048					
	934	N	101-200	SA 4	
12064					MAG SHARED 3C449
	5092	A	301-400	SN 8	
12066	MULT VEL				
	5700	E	201-300	AD 5	COMP VEL B
	5568	E	201-300	HS	COMP VEL B

UGC	NGC	R.A. (1950) DEC	m_{pg}	Hubble Type	Char.	Cluster	Group
12082		22 31.9 +32 37	15.6	SC-I			
12098	N7316	22 33.5 +20 04	13.7	S...	MARK	473	
12099	N7318	22 33.7 +33 43	14.9	E		471	12101
12100	N7318	22 33.7 +33 43	14.4	SB		471	12101
12101	N7320	22 33.8 +33 41	13.8	SC		471 BR IN GRP	12101
12102	N7319	22 33.8 +33 43	14.8	SB		471	12101
12113	N7331	22 34.8 +34 10	10.4	SB		471	
12115	N7332	22 35.0 +23 32	12.0	SO E			
12116	N7335	22 35.0 +34 12	14.7	SO-A SA		471 BR IN GRP	12116
12122	N7339	22 35.4 +23 32	13.1	SB-C			
12127		22 36.2 +35 05	15.0	E		471 BR IN GRP	12127
12129	N7343	22 36.3 +33 48	14.3	SB E		471	
12134		22 37.1 +11 31	14.9	SB-C SC			
12137		22 37.6 +37 57	14.0	SB-C SC		471	
12148	I5242	22 38.8 +23 08	14.7	COMP			
12153	I5243	22 39.0 +23 07	14.3	DBLE COMP			
12163		22 40.3 +29 28	14.4	S...			
12181		22 43.9 +37 48	14.5	SC	DISR		
12202		22 47.0 +11 03	14.8	SA		476	
12207	N7385	22 47.4 +11 21	14.4	E		476 BR IN CLUS	
12209	N7386	22 47.6 +11 25	14.6	E-SO SO		476	

UGC	V_r (km s^{-1})	Spectrum	Disp. (A/mm)	Ref.	Notes
12082	MULT VEL				
	806	N	21-CM	FT	
	815	N	21-CM	BA 4	
12098	MULT VEL				
	5700	E	201-300	AD 5	
	5469	E	201-300	HS	
12099	MULT VEL				
	6677	E	301-400	BB17	
	6724	A	401-500	L	
	6638	A	401-500	MW	
12100	MULT VEL				
	5638	A	401-500	MW	
	5867	A	401-500	L	
12101	MULT VEL				
	795	E	301-400	BB17	
	1168	N	301-400	V 8	
	750	N	21-CM	AL 1	
	755	N	21-CM	BA 3	
	774	N	21-CM	S 2	
12102	MULT VEL				
	6657	E	401-500	MW	
	6590	N	21-CM	BA 3	
	6620	N	21-CM	S 2	
12113	MULT VEL				
	780	E	201-300	MW	
	919	A	401-500	L	
	863	A	301-400	LB	
	850	E	301-400	BB36	
	815	N	21-CM	GU	
	815	N	21-CM	S 2	
	852	N	21-CM	R 3	
12115	MULT VEL				
	871	A	301-400	V 5	
	1204	A	401-500	MW	
	1190	A	1-100	MC	
	1200	N	21-CM	BI 1	
12116					
	6298	A	401-500	MW	
12122					
	1271	E	301-400	V 5	
12127	MULT VEL				
	8250	A	101-200	PT	
	8310	A	NO INFO	UL17	
12129					
	1216	A	401-500	MW	
12134					
	7318	E	1-100	FR11	
12137					
	4694	E	101-200	FR11	
12148					
	7173	E	301-400	KR	
12153	MULT VEL				
	7106	A	301-400	SA 2	
	7048	E	301-400	KR	
12163					
	7620	E	NO INFO	UL17	
12181					
	4787	E	101-200	FR11	
12202					
	7840	A	201-300	SR	
12207	MULT VEL				PKS2247+11
	7829	A	401-500	MW	
	7770	A	201-300	UL16	
	7764	A	201-300	SR	
12209	MULT VEL				
	7198	A	401-500	MW	
	7290	A	101-200	T	
	7317	A	201-300	SR	

UGC	NGC	R.A. (1950) DEC	m_{pg}	Hubble Type	Char.	Cluster	Group
12222		22 50.0 +11 23	15.2	SB		476	
12243		22 52.2 +11 27	15.6	SA-B		476	
12262	N7428	22 54.7 -01 18	13.8	SA			
12271		22 55.8 +02 02	14.9	SC			
12276	N7440	22 56.2 +35 32	14.6	SA			
12294	N7448	22 57.6 +15 42	12.0	SC		BR IN GRP	12294
12306	N7457	22 58.6 +29 53	12.3	SO			
12312	N7460	22 59.2 +02 00	14.2	SB SC	DISR		
12315	N7464	22 59.3 +15 42	14.5	... E			12294
12316	N7463	22 59.3 +15 43	13.5	S...	DISR		12294
12317	N7465	22 59.5 +15 42	13.3	SO	MARK		12294
12323		23 00.2 +32 20	15.0	SC			
12329	N7468	23 00.5 +16 20	14.0	PEC	MARK		
12332	N7469	23 00.7 +08 35	13.0	SA	SEYF	481	
12343	N7479	23 02.4 +12 03	11.7	SB			
12365	I5285	23 04.4 +22 40	14.4	COMP			
12391	N7495	23 06.5 +11 48	14.7	SC			
12397	N7499	23 07.8 +07 18	15.0	SO E		482	
12407		23 09.3 +09 14	14.2	S...		487	
12416		23 10.2 +10 28	15.4	PEC	MARK		
12417	I1474	23 10.3 +05 31	14.9	SC		487	
12422	N7518	23 10.6 +06 03	14.5	SA	MARK	487	
12423		23 10.6 +06 08	14.8	SC		487	

UGC	V_r (km s^{-1})	Spectrum	Disp. (A/mm)	Ref.	Notes
12222	8622	A	201-300	SR	
12243	8240	A	201-300	SR	
12262	3020	A	101-200	SN 8	
12271	MULT VEL				
	4767	E	101-200	SN 8	
	4987	E	1-100	FR11	
12276	5664	E	1-100	FR11	
12294	MULT VEL				
	2419	A	701-1000	MW	
	2247	N	21-CM	BG 3	
12306	MULT VEL				
	525	A	701-1000	MW	
	932	A	301-400	V 8	
12312	3296	E	101-200	SN 8	
12315	1872	E	301-400	V 5	
12316	2440	E	301-400	V 5	
12317	MULT VEL				
	1994	E	301-400	V 5	
	2100	E	201-300	AD 5	
	1777	E	101-200	HS	
12323	5965	E	101-200	FR11	
12329	MULT VEL				
	1800	E	201-300	AD 5	
	2208	E	301-400	BG 6	
	1991	E	101-200	HS	
	2093	N	21-CM	BG 6	
12332	MULT VEL				
	4868	E	101-200	BB26	
	4780	E	201-300	MW	
	4692	E	401-500	L	
	4850	E	1-100	UL12	
	4910	E	1-100	AN 2	
	4861	E	101-200	BW	
12343	MULT VEL				
	2421	E	301-400	BB10	
	2492	E	401-500	MW	
	2425	E	401-500	L	
12365	6221	A	301-400	SA 2	
12391	4895	E	101-200	FR11	4CP07.61
12397	MULT VEL				
	11916	A	401-500	MW	
	11595	N	101-200	TU	
12407	6574	E	NO INFO	CR 5	
12416	6900	E	301-400	DS 4	
12417	3371	A	NO INFO	CR 5	
12422	MULT VEL				
	3450	E	301-400	DS 4	
	643	N	101-200	TU	
	3569	E	NO INFO	CR 5	
					DISCREPANT VELOCITIES?
12423	MULT VEL				
	885	N	101-200	TU	
	4813	A	NO INFO	CR 5	
					DISCREPANT VELOCITIES?

UGC	NGC	R.A. (1950) DEC	m_{pg}	Hubble Type	Char.	Cluster	Group
12431	N7529	23 11.5 +08 42	14.6	S...		487	
12442	N7537	23 12.0 +04 13	13.8	SB SB-C		487	
12447	N7541	23 12.2 +04 15	12.7	SC		487	
12453	N7547	23 12.5 +18 42	14.9	SA		486	12456
12454		23 12.6 +09 24	15.0	SO		487	
12456	N7550	23 12.8 +18 41	13.9	E-SO		486 BR IN GRP	12456
12457	N7549	23 12.8 +18 46	14.1	S...		486	12456
12462		23 13.3 +09 14	15.5	SB		487	
12464	N7562	23 13.4 +06 24	13.0	E		487	
12472		23 14.2 +08 37	14.8	SO		487	
12477	N7578	23 14.7 +18 25	15.0	DBLE		486	
12478	N7578	23 14.8 +18 26	15.0	DBLE		486	
12481	N7580	23 15.1 +13 44	14.8	SB	MARK		
12484	N7587	23 15.5 +09 24	14.9	SA		487	
12486	N7591	23 15.8 +06 18	13.8	SB		487	
12487	N7601	23 16.2 +08 57	14.7	SB-C		487	
12490		23 16.2 +24 57	14.0	SA	MARK		
12493	N7603	23 16.4 -00 01	14.4	S...	SEYF	485	
12494		23 16.4 +06 35	15.0	SC		487	
12498	I5309	23 16.6 +07 50	15.0	SB			
12500	N7608	23 16.7 +08 04	15.2	S...		487	
12509	N7611	23 17.1 +07 46	14.0	SO		487	
12511	N7610	23 17.1 +09 54	14.9	SC	DISR	487	
12512	N7612	23 17.2 +08 17	14.3	SO		487	
12520	N7620	23 17.6 +23 56	13.5	SC	MARK		
12523	N7619	23 17.7 +07 55	12.7	E		487	
12526	N7623	23 17.9 +08 06	13.9	SO E		487	
12527	N7624	23 17.9 +27 03	13.7	SC	MARK		

UGC	V_r (km s^{-1})	Spectrum	Disp. (A/mm)	Ref.	Notes
12431					
	4570	E	NO INFO	CR 5	
12442	MULT VEL				
	2682	E	301–400	V 5	
	2601	N	101–200	TU	
	2711	E	NO INFO	CR 5	
12447	MULT VEL				
	2672	A	401–500	MW	
	2392	N	101–200	TU	
12453					
	4858	E	201–300	CR 1	
12454					MAG SHARED
	4758	A	NO INFO	CR 5	
12456					
	4987	E	201–300	CR 2	
12457					
	4806	E	201–300	CR 1	
12462					
	12513	A	NO INFO	CR 5	
12464	MULT VEL				
	3806	A	401–500	L	
	3523	N	101–200	TU	
12472					
	6513	E	NO INFO	CR 5	
12477					MAG SHARED
	11952	A	201–300	CR 2	
12478					MAG SHARED
	12093	A	201–300	CR 2	
12481	MULT VEL				
	4800	E	201–300	AD 5	
	4732	E	101–200	HS	
12484					
	8934	A	NO INFO	CR 5	
12486					
	5002	E	NO INFO	CR 5	
12487					
	8207	E	NO INFO	CR 5	
12490	MULT VEL				
	8400	E	201–300	AD 5	
	8051	E	301–400	KR	
12493	MULT VEL				
	8850	E	301–400	DS 4	
	8700	E	101–200	TO	
	8800	E	101–200	A 6	
	8700	E	101–200	DS 4	
12494					MAG SHARED
	4238	E	NO INFO	CR 5	
12498					
	4266	E	NO INFO	CR 5	
12500					
	3561	E	NO INFO	CR 5	
12509	MULT VEL				
	3383	A	401–500	MW	
	3303	A	NO INFO	CR 5	
12511					
	3506	N	101–200	FR11	
12512					
	3189	A	NO INFO	CR 5	
12520					
	9600	E	201–300	AD 5	
12523	MULT VEL				
	3757	A	401–500	MW	
	3990	N	NO INFO	SN 5	
12526					
	3463	A	401–500	MW	
12527	MULT VEL				
	4500	E	201–300	AD 5	
	4188	N	21–CM	BI 1	

UGC	NGC	R.A. (1950) DEC	m_{pg}	Hubble Type	Char.	Cluster	Group
12529	N7625	23 18.0 +16 57	12.8	••• S0		486	
12531	N7626	23 18.2 +07 56	12.8	E		487	
12539	N7631	23 18.8 +07 56	13.8	SB		487	
12542	N7634	23 19.1 +08 36	13.7	S0		487 BR IN GRP	12542
12554	N7640	23 19.7 +40 35	11.6	SC			
12570		23 20.7 +32 15	14.5	PEC		490	
12575	N7648	23 21.3 +09 23	13.5	S0	MARK	487	
12579	N7649	23 21.8 +14 22	15.7	E		489	
12602	N7671	23 24.8 +12 12	14.3	S0		487	
12607	N7673	23 25.2 +23 19	12.7	COMP	MARK	493	
12608	N7674	23 25.4 +08 30	13.6	SB	MARK	487	
12610	N7677	23 25.6 +23 15	13.9	SB	MARK	493	
12613		23 26.0 +14 27	15.5	IRR			
12614	N7678	23 26.0 +22 08	12.7	SC		493	
12618	N7679	23 26.2 +03 14	13.2	S0	MARK		
12620		23 26.4 +17 03	14.2	S0	DISR		
12622	N7682	23 26.5 +03 15	14.3	SA	DISR		
12632		23 27.6 +40 43	15.7	SC-I			

UGC	V_r (km s^{-1})	Spectrum	Disp. (Å/mm)	Ref.	Notes
12529	MULT VEL				
	1706	E	401-500	MW	
	1828	E	401-500	L	
	1620	E	101-200	UL 5	
	1641	N	21-CM	BA 2	
	1620	N	21-CM	PS	
	1603	N	21-CM	BI 1	
	1654	N	21-CM	CX	
12531	MULT VEL				PKS2318+07
	3613	A	1-100	DC	
	3357	A	401-500	MW	
12539					
	3773	A	NO INFO	CR 5	
12542	MULT VEL				
	3190	A	201-300	BC	
	3236	A	NO INFO	CR 5	
12554	MULT VEL				
	388	E	301-400	V 5	
	423	E	401-500	L	
	373	N	21-CM	DD	
	370	N	21-CM	RR	
	380	N	21-CM	GU	
	363	N	21-CM	R 3	
12570					
	5310	E	NO INFO	UL17	
12575	MULT VEL				
	4050	E	301-400	DS 4	
	3593	E	NO INFO	CR 5	
12579					DISCREPANT VELOCITIES?
	12514	A	301-400	MW	
12602	MULT VEL				
	4129	A	401-500	L	
	4217	N	101-200	TU	
	4126	N	NO INFO	CR 5	
12607	MULT VEL				
	3055	E	301-400	SA 2	
	3300	E	201-300	AD 5	
	3391	E	301-400	KR	
	3402	E	301-400	BG 6	
	3404	N	21-CM	BG 6	
	3359	N	21-CM	BI 1	
	3408	N	21-CM	CX	
12608	MULT VEL				MAG SHARED
	8850	E	301-400	DS 4	
	8709	E	NO INFO	CR 5	
12610	MULT VEL				
	3900	E	201-300	AD 5	
	3637	E	301-400	KR	
	3583	E	201-300	BG 6	
	3540	N	21-CM	BG 6	
	3489	N	21-CM	BI 1	
12613					
	-179	N	21-CM	FT	
12614					
	3446	E	401-500	MW	
12618	MULT VEL				
	5202	E	401-500	MW	
	5101	E	401-500	L	
	5203	N	101-200	TU	
	5250	E	301-400	DS 4	
	5174	E	NO INFO	BB38	
12620					
	6992	N	101-200	TU	
12622					
	5168	N	101-200	TU	
12632					
	424	N	21-CM	FT	

UGC	NGC	R.A. (1950) DEC	m pg	Hubble Type	Char.	Cluster	Group
12665		23 31.2 +29 46	15.1	S•••			
12682		23 32.4 +17 57	14.7	IRR			
12699	N7714	23 33.7 +01 53	13.1	S•••	DISR		
12700	N7715	23 33.8 +01 53	14.9	S••• SC			
12702	N7716	23 34.0 +00 02	12.9	SB			
12703	I5338	23 34.0 +20 52	15.5	E		492	
12709		23 34.9 +00 08	15.7	SC-I			
12713		23 35.7 +30 25	15.1	•••			
12716	N7720	23 36.0 +26 45	13.9	DBLE E		493	
12719		23 36.2 +26 30	15.7	SO-A		493	
12721		23 36.7 +26 50	15.0	SB		493	
12727	N7728	23 37.4 +26 52	14.3	E		493	
12733		23 38.2 +26 33	15.1	E		493	
12737	N7731	23 39.0 +03 28	14.3	SA			
12738	N7732	23 39.1 +03 27	14.5	SC-I	DISR		
12747		23 40.5 +19 09	14.6	IRR	MARK		
12754	N7741	23 41.3 +25 48	11.8	SC		493	
12759	N7743	23 41.8 +09 39	12.9	SO-A SA		494	
12760	N7742	23 41.8 +10 30	12.5	SO SB			
12767		23 42.6 +06 46	14.8	SB SC		495	
12777	N7750	23 44.1 +03 30	13.8	SC			
12779	N7752	23 44.5 +29 10	14.3	S•••			
12780	N7753	23 44.6 +29 12	13.2	SB SC			
12788	N7757	23 46.2 +03 55	13.9	SC			

UGC	V_r (km s^{-1})	Spectrum	Disp. (A/mm)	Ref.	Notes
12665					
	4900	E	101-200	KZ	
12682					
	1394	N	21-CM	FT	
12699	MULT VEL				
	2833	E	401-500	L	
	2818	E	301-400	KR	
	2780	E	101-200	UL 2	
	2805	N	21-CM	PS	
12700					
	2795	A	401-500	L	
12702					
	2546	A	401-500	MW	
12703					
	16620	A	101-200	SA 4	
12709					
	2677	N	21-CM	FT	
12713					
	240	E	NO INFO	UL17	
12716	MULT VEL				3C465
	9150	E	101-200	PT	
	8790	E	301-400	ST	
	8485	A	201-300	J	
	9045	A	201-300	SR	COMP VEL C
	7911	A	201-300	SR	COMP VEL N
					DISCREPANT VELOCITIES?
12719					
	9726	A	201-300	SR	
12721					
	7499	A	201-300	SR	
12727	MULT VEL				2337+26W1
	9420	A	201-300	UL16	
	9534	A	201-300	SR	
12733					
	11560	A	201-300	SR	
12737	MULT VEL				
	2800	N	101-200	TU	
	2850	A	NO INFO	Z 3	
12738	MULT VEL				
	2968	E	301-400	KR	
	2873	N	101-200	TU	
12747	MULT VEL				
	4200	E	201-300	AD 5	
	4037	E	201-300	HS	
12754	MULT VEL				
	808	E	301-400	V 5	
	729	A	401-500	MW	
	718	E	301-400	V 8	
	786	N	21-CM	R 3	
12759					
	1802	E	401-500	MW	
12760	MULT VEL				
	1629	E	401-500	MW	
	1748	E	401-500	L	
12767					
	5324	E	101-200	SN 8	
12777					
	2883	E	101-200	SN 8	
12779	MULT VEL				
	4868	A	301-400	LB	
	4855	E	1-100	A 4	
	4989	E	301-400	KR	
12780	MULT VEL				
	4845	E	301-400	LB	
	5423	E	301-400	KR	
	5206	E	1-100	BO	
12788					
	3118	E	101-200	SN 8	

	UGC	NGC	R.A. (1950) DEC	m_{pg}	Hubble Type	Char.	Cluster	Group
	12791		23 46.3 +25 57	15.2	IRR		493	
	12806	N7768	23 48.4 +26 53	14.0	E		493	
174	12808	N7769	23 48.5 +19 52	12.9	SA-B SC		BR IN	12808 GRP
	12810		23 48.6 +00 46	14.4	SB SC			
	12813	N7770	23 48.8 +19 49	14.5	... SB	DISR		12808
	12815	N7771	23 48.9 +19 50	13.1	SA SB	DISR		12808
	12833	N7780	23 51.0 +07 50	14.8	SA-B SC		495	
	12834	N7782	23 51.3 +07 42	13.2	SB SC		BR IN	12834 GRP
	12837	N7783	23 51.6 +00 06	14.1	DBLE SO-A		BR IN	12837 GRP
	12841	N7785	23 52.8 +05 38	13.0	S... E			
	12848	I1515	23 53.6 −01 16	14.8	SB SC	DISR		
	12852	I1516	23 53.7 −01 12	14.3	SB-C SC			
	12884	N7798	23 56.9 +20 29	12.7	S...	MARK		
	12914		23 59.1 +23 13	13.2	S...	DISR		
	12915		23 59.2 +23 14	13.9	...	DISR		

UGC	V_r (km s^{-1})	Spectrum	Disp. (A/mm)	Ref.	Notes
12791					
	799	N	21-CM	FT	
12806	MULT VEL				
	7950	A	101-200	PT	
	7143	A	201-300	MS	DISCREPANT VELOCITIES?
12808					
	4349	E	401-500	L	
12810					
	8214	E	101-200	SN 8	
12813					
	4338	E	401-500	L	
12815	MULT VEL				
	4276	E	401-500	L	
	4293	A	301-400	L B	
12833					
	5125	A	101-200	SN 8	
12834	MULT VEL				
	5338	A	101-200	SN 8	
	5960	A	301-400	V 8	DISCREPANT VELOCITIES?
12837	MULT VEL				
	7012	N	101-200	TU	COMP VEL A
	6702	N	101-200	TU	COMP VEL B
12841					
	3846	A	401-500	MW	
12848					
	6709	E	101-200	SN 8	
12852					
	7360	A	101-200	SN 8	
12884					
	2700	E	201-300	AD 5	
12914	MULT VEL				
	4581	E	301-400	KR	
	4591	N	NO INFO	BB29	
12915	MULT VEL				
	4313	E	301-400	KR	
	4585	N	NO INFO	BB29	

175

Appendix

Uppsala Labels

for Zwicky "Near" Distance Class Clusters,

and Identifications with Abell Numbers

and Common Names

CLUS	POSITION	CHARACTER	POP	RADIUS	CORRESPONDING CLUSTERS
1	0000.8+0452	OPEN	157	12.1	
2	0002.4+0744	OPEN	177	14.0	
3	0013.4+1805	OPEN	135	12.7	
4	0013.6+2927	MED COMP	145	9.5	
5	0014.5+2315	OPEN	245	23.6	
6	0024.4+3014	OPEN	144	16.3	
7	0032.6+0207	OPEN	170	16.0	
8	0033.8+0538	MED COMP	238	14.7	A 76
9	0034.4+2532	OPEN	101	6.0	
10	0036.3+2914	OPEN	118	8.5	A 71
12	0046.5+2300	MED COMP	106	6.2	
14	0054.6−0127	COMPACT	278	9.0	A 119
15	0055.0+1212	OPEN	107	9.2	
16	0056.9+2636	MED COMP	145	7.0	
17	0103.7+3942	MED COMP	163	9.1	
18	0105.5+3650	OPEN	190	19.7	
19	0106.9+0028	OPEN	225	17.8	A 147
20	0107.5+3212	MED COMP	625	32.0	
21	0110.5+1515	MED COMP	254	10.3	
22	0119.6+5035	OPEN	316	21.4	
23	0121.5+0113	OPEN	160	12.8	
24	0123.0+0953	OPEN	167	12.5	
25	0123.6−0133	MED COMP	262	20.2	A 194
26	0127.6+1828	MED COMP	227	18.8	A 195
28	0143.8+2323	MED COMP	174	11.4	
29	0144.0+1230	OPEN	135	16.4	
30	0145.8+4740	OPEN	265	16.7	
31	0150.8+3615	MED COMP ·	450	18.9	A 262
32	0150.9+3050	MED COMP	570	27.8	A 278
33	0202.6+1852	OPEN	146	12.2	
34	0205.5+0110	OPEN	131	11.6	
35	0208.0+1515	OPEN	430	27.0	
36	0210.9+5038	OPEN	182	11.9	
37	0216.0+3625	MED COMP	485	24.1	
38	0226.0+2600	OPEN	1184	50.0	
39	0233.0+0124	MED COMP	180	14.7	
40	0236.2+3249	MED COMP	211	18.0	
41	0240.6+0740	OPEN	290	20.8	
42	0241.2+3558	MED COMP	313	13.5	A 376
43	0246.1−0045	OPEN	130	13.3	
44	0248.0+1307	MED COMP	223	13.8	
45	0254.7+0555	MED COMP	194	14.0	A 400
46	0254.7+1606	MED COMP	183	10.5	A 397
47	0303.0+4125	MED COMP	5130	77.1	PERSEUS A 347 A 407 A 426

CLUS	POSITION	CHARACTER	POP	RADIUS	CORRESPONDING CLUSTERS
48	0307.7+1907	MED COMP	117	10.5	
49	0310.0-0130	OPEN	540	30.0	
50	0312.2+1551	OPEN	204	14.0	
51	0321.0+0648	OPEN	129	16.1	
52	0337.8+1536	MED COMP	157	10.0	
53	0354.0+7900	OPEN	205	13.9	
55	0403.1+3040	OPEN	175	18.4	
56	0414.5+0216	OPEN	225	13.9	
57	0417.7+3557	OPEN	131	11.2	
59	0426.4-0002	OPEN	155	17.3	
59A	0430.8-0424	OPEN	385	21.9	
62	0444.7+0828	OPEN	231	10.3	
62A	0449.3-0437	OPEN	268	22.2	
63	0451.3+0159	OPEN	125	8.4	
64	0452.2+7305	OPEN	230	19.3	
65	0456.1-0103	MED COMP	144	13.6	
66	0506.6+1649	OPEN	151	13.4	
67	0507.9-0110	MED COMP	140	8.1	
68	0510.0+0458	COMPACT	1071	31.7	A 539
69	0517.8+5108	MED COMP	108	8.4	
70	0521.2+6418	OPEN	347	22.8	
71	0529.2+0516	OPEN	153	17.0	
72	0531.3+0127	OPEN	157	14.4	
74	0544.4+5036	OPEN	230	17.7	
75	0558.5+5951	OPEN	310	24.5	
76	0603.0+7922	OPEN	151	15.5	
80	0628.9+5232	OPEN	785	28.6	
81	0631.6+2609	MED COMP	140	6.5	
83	0632.7+6323	MED COMP	144	7.3	
84	0640.4+2559	OPEN	155	9.2	
85	0642.0+7334	OPEN	155	12.8	
86	0642.2+4130	OPEN	569	32.9	
87	0647.4+3323	OPEN	167	7.5	
88	0654.8+2753	OPEN	397	17.4	
89	0700.4+4801	MED COMP	1273	36.4	A 569
90	0701.0+1858	MED COMP	129	8.0	
91	0705.4+3642	OPEN	236	12.7	
92	0706.6+3221	OPEN	360	11.2	
94	0710.5+4222	MED COMP	575	18.6	
95	0718.6+3249	OPEN	362	14.7	
96	0718.9+5412	MED COMP	1450	28.9	A 576
97	0720.6+2259	OPEN	217	14.2	
98	0730.1+1858	MED COMP	452	15.6	
99	0731.9+3125	OPEN	178	8.0	
100	0733.4+6102	MED COMP	1315	36.3	
101	0733.8+2356	OPEN	263	9.3	
102	0735.0+8545	OPEN	216	16.9	
103	0739.8+4949	MED COMP	250	15.3	
105	0744.3+1839	MED COMP	203	9.0	
106	0745.5+4020	OPEN	212	16.7	
107	0746.5+2315	OPEN	172	10.0	
108	0752.9+2833	OPEN	933	22.8	
110	0754.0+4619	OPEN	125	10.0	
111	0755.9+0805	MED COMP	90	4.0	
112	0756.1+5616	OPEN	430	16.8	
113	0800.0+0946	OPEN	207	8.5	
114	0801.3+3954	OPEN	144	9.5	
115	0806.8+0514	MED COMP	220	6.5	
116	0810.1+5813	MED COMP	141	8.6	
118	0819.6+2209	MED COMP	325	21.6	CANCER A 634
119	0820.1-0029	OPEN	90	6.5	
120	0820.6+0436	MED COMP	163	5.5	
121	0822.4+5453	OPEN	320	14.5	
122	0824.8+1731	OPEN	355	9.8	
123	0826.2+3039	COMPACT	778	13.5	A 671
124	0829.6+5245	MED COMP	128	6.5	
125	0832.2+3845	MED COMP	130	5.5	
126	0832.6-0235	OPEN	490	17.4	
127	0836.3+4147	OPEN	550	27.6	
128	0837.0+2506	OPEN	403	12.7	
129	0837.1+6445	OPEN	135	14.6	
130	0842.3+2820	OPEN	280	7.6	
131	0846.3+1910	MED COMP	267	10.4	
132	0851.1+4925	MED COMP	110	6.0	
133	0852.7+0054	OPEN	163	4.4	
134	0853.5-0312	OPEN	189	10.4	
135	0855.0+5248	OPEN	295	18.0	
136	0856.3+4554	OPEN	105	6.8	
138	0901.2+6640	OPEN	189	16.7	
139	0903.9+3716	OPEN	147	9.4	
140	0909.7+1814	OPEN	1306	31.7	
141	0911.0+3025	OPEN	190	9.8	
142	0915.6+3409	MED COMP	185	8.4	A 779
143	0916.7+4952	OPEN	199	15.7	
145	0920.4+4037	OPEN	108	5.2	
146	0921.6+2354	MED COMP	636	16.2	
147	0923.9+7353	MED COMP	348	11.7	
148	0926.5+3026	OPEN	371	10.5	
149	0927.2+3446	OPEN	173	10.9	

CLUS	POSITION	CHARACTER	POP	RADIUS	CORRESPONDING CLUSTERS
150	0935.3+1701	MED COMP	177	7.8	
151	0938.2+1130	OPEN	196	7.5	
152	0938.7+3957	OPEN	110	7.8	
153	0941.6+1540	OPEN	142	7.0	
154	0941.7+2430	OPEN	770	33.6	
155	0943.7+5454	MED COMP	187	8.8	
156	0956.4+3730	OPEN	347	14.4	
157	0957.9+1111	OPEN	290	9.0	
158	0958.9+0038	MED COMP	387	9.3	
159	1003.6+1443	OPEN	306	8.4	
161	1010.5+3922	MED COMP	160	6.0	
162	1012.0-0047	MED COMP	209	5.2	
163	1012.8+5337	OPEN	685	28.3	
164	1014.1+2215	OPEN	194	9.9	
166	1019.4+2501	MED COMP	266	8.9	
166A	1019.5+0041	OPEN	169	6.5	
167	1019.8+3653	MED COMP	456	18.3	
168	1020.1+1306	OPEN	229	6.0	A 999
169	1020.1+2046	OPEN	256	8.1	
170	1020.4-0316	MED COMP	1150	21.8	
171	1020.4+4210	OPEN	142	6.4	
172	1021.0+7728	OPEN	765	32.2	
173	1021.8+1725	MED COMP	318	7.8	
174	1023.8+1056	MED COMP	232	6.0	A1016
175	1026.1+0412	MED COMP	634	6.3	
176	1026.2+2215	OPEN	248	8.7	
177	1026.9+4023	MED COMP	570	13.0	A1035
179	1028.2+4357	OPEN	200	10.3	
180	1029.3+5736	MED COMP	210	12.0	
181	1029.8+2023	OPEN	226	7.4	
182	1033.3+3804	OPEN	198	7.0	
183	1034.9+2912	OPEN	209	9.2	
184	1035.1+7226	OPEN	235	16.7	
185	1037.4+2156	OPEN	170	9.4	
186	1039.3+1109	OPEN	244	7.9	
187	1039.4+1649	MED COMP	164	5.7	
188	1042.4+3910	MED COMP	344	12.7	
189	1043.9+5635	OPEN	74	5.5	
190	1045.8+0510	MED COMP	241	4.0	
191	1046.3+0038	MED COMP	127	4.6	
192	1046.8+2747	OPEN	412	15.4	
193	1047.6+1623	MED COMP	137	4.5	
194	1048.6+2358	OPEN	770	17.5	
195	1049.2+0902	OPEN	172	5.5	
196	1055.0+1725	OPEN	122	5.7	
197	1055.4+0142	MED COMP	224	6.5	A1139
198	1056.1+0644	OPEN	353	5.6	
199	1056.5+1240	OPEN	150	6.7	
200	1056.9+0922	MED COMP	479	10.0	
201	1057.2+3804	OPEN	535	13.2	
202	1058.6+1049	MED COMP	334	5.0	A1142
203	1058.6+4611	OPEN	140	11.2	
204	1102.1+3136	MED COMP	281	8.8	
205	1105.2+1342	OPEN	165	6.4	
206	1105.3+2835	MED COMP	1090	21.0	A1185
207	1105.6+2323	MED COMP	723	11.7	
208	1106.2+0516	MED COMP	540	10.6	
209	1107.6+1041	OPEN	333	8.0	
210	1107.7+3610	OPEN	293	13.9	
212	1112.7+7259	OPEN	136	10.7	
213	1112.9+2600	OPEN	98	5.5	
214	1114.3+5457	MED COMP	245	11.2	A1225
215	1115.2+3013	MED COMP	950	15.5	A1213
215A	1116.0-0410	OPEN	1763	28.7	
216	1117.0+4653	OPEN	210	12.8	
217	1117.6+3352	OPEN	290	10.0	A1228
218	1119.7+0305	MED COMP	510	10.0	
219	1122.3+6317	OPEN	156	13.1	
220	1123.5+2256	OPEN	350	14.9	
221	1123.9+3541	OPEN	260	11.0	A1257
222	1125.5+2759	OPEN	673	13.7	A1267
223	1126.3+0913	OPEN	213	6.5	
225	1130.9+3435	OPEN	166	4.2	
226	1131.2+4923	COMPACT	257	7.9	A1314
227	1131.6+4421	OPEN	113	7.3	
228	1138.3+1024	MED COMP	138	6.0	
229	1138.7+5650	MED COMP	2690	49.2	A1270 A1291 A1318 A143
230	1140.0+2715	OPEN	247	8.2	
231	1141.7-0158	OPEN	2320	33.2	
233	1141.9+3004	OPEN	214	7.8	
234	1142.1+2126	MED COMP	1895	23.8	A1367
235	1142.2+3456	MED COMP	470	19.4	
238	1146.8+1237	MED COMP	195	3.8	
239	1148.6+0642	OPEN	119	3.7	
242	1153.0+2522	MED COMP	330	12.5	
243	1154.9+2806	OPEN	209	6.3	
244	1155.0+3127	MED COMP	209	8.6	
245	1156.2+2201	OPEN	188	7.9	
248	1157.2+4332	OPEN	463	15.5	

CLUS	POSITION	CHARACTER	POP	RADIUS	CORRESPONDING CLUSTERS	
249	1201.3+0151	MED COMP	220	6.7		
250	1201.5+3916	MED COMP	210	8.9		
251	1202.0+2028	OPEN	555	9.7		
252	1202.9+1051	OPEN	272	7.0		
253	1204.7+2246	OPEN	178	6.7		
254	1204.7+3319	OPEN	123	5.5		
255	1204.8+6520	OPEN	515	26.7		
256	1205.4+2515	MED COMP	297	9.4		
257	1207.6+5235	OPEN	161	9.0		
258	1208.1+3603	OPEN	180	12.2		
259	1211.4+6013	MED COMP	220	10.3		
260	1212.0+2409	OPEN	139	9.5		
261	1216.5+6148	OPEN	135	10.2		
263	1217.5+2915	MED COMP	1828	30.9		
265	1222.6+4548	MED COMP	162	10.7		
266	1224.1+0914	MED COMP	244	2.4		
267	1224.6+3131	MED COMP	521	11.9		
268	1226.0+0859	MED COMP	61	1.3		
269	1230.3+7450	OPEN	207	17.6		A1500
270	1231.4+5044	MED COMP	293	7.8		
271	1231.8+4810	OPEN	175	9.0		
273	1243.2+4143	OPEN	139	10.6		
274	1245.0-0002	MED COMP	222	7.0		
275	1255.6+7858	OPEN	141	12.0		
276	1257.1+2806	COMPACT	2150	28.5	COMA	A1656
277	1257.3+3925	MED COMP	160	6.8		
278	1301.9+5001	OPEN	235	11.6		
279	1302.2+6243	OPEN	174	14.0		
280	1307.0+3944	COMPACT	355	8.6		A1691
281	1308.2+3531	OPEN	210	14.3		
282	1308.3+4456	MED COMP	477	20.3		
283	1309.3+2255	OPEN	259	16.0		
284	1312.3+4115	OPEN	208	9.2		
285	1313.0+5410	MED COMP	623	27.3		
286	1317.9+3350	COMPACT	147	8.4		
287	1319.6+3135	OPEN	411	16.2		
288	1321.4+1358	OPEN	278	8.0		
290	1327.3+1145	OPEN	449	10.5		
291	1330.7+1347	MED COMP	373	8.5		
292	1338.4+2420	MED COMP	320	9.3		
293	1339.9+3030	MED COMP	417	11.2		A1781
294	1341.0+5930	OPEN	560	35.3		
295	1341.6+2614	MED COMP	694	12.5		
297	1342.5+7841	OPEN	135	11.7		
298	1343.3+2245	OPEN	126	6.1		
299	1346.2+2814	COMPACT	320	8.4		A1800
300	1347.5+1815	OPEN	455	12.5		
301	1348.7+0249	OPEN	166	7.3		
303	1350.9+2142	MED COMP	214	8.5		
304	1351.3+3333	MED COMP	129	6.2		
305	1352.0+3107	OPEN	105	5.4		
306	1352.9+3856	MED COMP	310	22.1		
307	1353.2+2508	MED COMP	240	12.7		
308	1353.7+0553	OPEN	290	11.2		
309	1355.1+2237	MED COMP	310	7.8		
310	1357.1+2836	COMPACT	790	13.1		A1831
311	1357.6+3244	OPEN	122	7.4		
312	1358.7+1521	OPEN	923	24.3		
313	1400.4+0949	MED COMP	192	12.0		
314	1404.8+1254	MED COMP	207	4.5		
316	1406.4+5513	OPEN	204	8.1		
317	1410.0+2509	OPEN	199	8.1		
318	1415.3+0038	OPEN	121	5.5		
319	1416.0+0752	MED COMP	720	8.0		A1890
320	1416.2+3606	OPEN	161	13.1		
322	1420.7+4025	MED COMP	145	6.2		
323	1422.0+1732	OPEN	803	11.6		
324	1424.0+2613	MED COMP	450	18.2		
325	1426.4+1132	OPEN	335	8.5		
326	1426.8+2947	OPEN	603	24.7		
327	1429.9+0336	MED COMP	127	4.5		
328	1429.9+5256	OPEN	190	14.9		
329	1431.9+6020	OPEN	178	13.4		
330	1436.0+0926	OPEN	920	16.5		
332	1438.4+0405	MED COMP	428	11.3		
333	1440.3+0128	OPEN	345	13.3		
334	1441.2+1909	OPEN	358	9.6		
335	1442.0+2414	MED COMP	1413	31.2		
336	1445.2+1356	OPEN	155	5.5		
337	1445.4+1125	OPEN	355	7.0		
339	1448.7+1651	COMPACT	583	11.0		A1983
340	1448.8+4654	MED COMP	143	7.0		
341	1451.6+1855	COMPACT	510	8.6		A1991
342	1454.3+0915	MED COMP	289	7.0		
343	1456.2+4901	OPEN	209	12.5		
344	1457.5+5415	OPEN	245	19.9		
345	1459.8+2043	OPEN	741	13.5		
346	1500.6+2559	OPEN	259	9.8		
349	1503.8+0853	OPEN	1093	11.5		

CLUS	POSITION	CHARACTER	POP	RADIUS	CORRESPONDING CLUSTERS
350	1504.6+1639	OPEN	232	6.8	
351	1508.8+4054	OPEN	200	17.9	
352	1510.0+0315	OPEN	2295	31.4	
355	1519.4+2610	OPEN	396	14.3	
357	1521.0+4844	OPEN	230	13.3	
358	1521.2+0851	COMPACT	640	9.0	A2063
359	1524.8+2850	COMPACT	984	13.4	A2079
360	1530.9+0454	MED COMP	211	6.5	
362	1534.0+4222	MED COMP	290	21.7	
363	1534.4+2553	OPEN	122	6.0	
364	1534.4+3100	OPEN	196	9.1	
365	1536.5+2147	MED COMP	293	12.4	
366	1539.1+2820	MED COMP	241	9.5	
367	1540.9+4758	OPEN	152	9.0	
368	1544.4+2543	MED COMP	240	11.2	
370	1546.0+6722	OPEN	169	12.9	
371	1548.6+2136	OPEN	182	11.3	
373	1549.6+2417	OPEN	157	8.7	
374	1550.4+1243	OPEN	341	8.3	
378	1555.1+4146	MED COMP	137	10.1	
379	1556.1+2829	OPEN	226	8.2	
381	1559.0+5353	COMPACT	281	9.4	
382	1600.4+1925	MED COMP	2859	33.8	HERCULES A2147 A2151 A2152
383	1600.9+2528	MED COMP	246	5.8	A2148
384	1601.5-0201	MED COMP	99	3.7	
385	1603.7+0006	OPEN	438	11.6	
386	1608.5+3044	MED COMP	1775	31.8	A2162
387	1609.0+6411	OPEN	167	10.3	
388	1609.0+8212	MED COMP	260	17.5	A2247
389	1610.3+4955	OPEN	675	26.7	
390	1610.7+1110	MED COMP	130	4.0	
391	1611.6+3717	OPEN	351	18.4	
392	1613.8+5632	OPEN	393	19.7	
393	1615.8+3505	MED COMP	172	10.6	
395	1625.5+4006	MED COMP	686	16.3	A2197 A2199
396	1626.2+2045	OPEN	324	11.0	
397	1626.6+3326	OPEN	75	7.8	
399	1629.7+5027	MED COMP	147	5.5	
400	1634.4+4412	MED COMP	127	7.0	
401	1635.5+2608	OPEN	390	15.8	
402	1635.9+2939	MED COMP	122	5.6	
403	1638.4+6038	OPEN	510	30.6	
405	1645.1+0620	OPEN	87	5.4	
406	1647.6+5337	MED COMP	157	10.0	
408	1648.6-0307	OPEN	195	9.3	
409	1649.9+2343	MED COMP	310	11.5	
411	1653.9+7856	COMPACT	225	12.5	A2256
412	1655.8+6844	OPEN	292	20.3	
414	1658.5+2459	MED COMP	244	11.6	
415	1701.4+2830	OPEN	340	16.2	
416	1702.9+3510	MED COMP	273	16.6	
419	1704.9+3056	MED COMP	191	10.5	
420	1707.6+4045	OPEN	1021	29.8	
421	1712.1+7740	MED COMP	153	8.1	A2248
422	1712.6+0616	OPEN	145	8.5	
423	1715.9+3218	MED COMP	101	6.3	
425	1718.6+5229	OPEN	230	14.1	
427	1728.5+4353	MED COMP	157	11.2	
429	1730.4+5829	OPEN	417	26.0	
430	1730.9+7304	MED COMP	345	20.2	
431	1738.0+3516	MED COMP	151	9.2	
432	1742.2+2345	MED COMP	93	10.0	
433	1743.8+5528	OPEN	128	6.8	
434	1744.5+3846	OPEN	205	16.8	
435	1745.6+6703	MED COMP	297	16.0	
436	1752.6+1842	OPEN	86	6.0	
437	1754.9+6230	MED COMP	140	8.5	
438	1756.5+2904	OPEN	224	20.3	
439	1759.9+4238	OPEN	155	11.1	
440	1807.2+5633	MED COMP	130	8.6	
441	1808.9+2531	OPEN	101	12.6	
442	1810.2+4949	MED COMP	149	10.8	
443	1811.4+6941	MED COMP	229	8.9	
444	1612.3+2237	MED COMP	210	18.5	
445	1822.2+3009	MED COMP	148	9.2	
446	1826.4+3410	MED COMP	163	10.8	
447	1831.2+3154	MED COMP	118	6.6	
448	1836.8+3306	MED COMP	77	5.2	
449	1837.1+3633	OPEN	165	12.5	
450	1847.2+7711	OPEN	325	14.6	
451	1853.0+7226	MED COMP	175	8.4	
453	1916.8+4855	MED COMP	3755	73.7	A2319
454	1954.8+7824	OPEN	206	16.4	
458	2044.9-0315	OPEN	563	30.3	
460	2100.0+1623	OPEN	170	12.3	
461	2110.2+1238	OPEN	215	18.7	
462	2110.9+0203	MED COMP	172	9.7	
463	2134.6+4253	OPEN	95	16.2	
464	2149.6+0319	OPEN	186	9.8	

CLUS	POSITION	CHARACTER	POP	RADIUS	CORRESPONDING CLUSTERS	
466	2202.7+1628	MED COMP	150	9.0		
467	2207.8+4114	OPEN	135	14.7		
468	2210.0+3745	MED COMP	173	15.6		
469	2212.0+1326	OPEN	136	7.9		
471	2231.2+3732	OPEN	782	31.8		
472	2231.5+0052	MED COMP	228	11.5		
473	2233.3+1911	OPEN	197	12.8		
474	2236.7+2442	OPEN	123	9.7		
475	2237.5-0100	OPEN	170	10.0		
476	2247.3+1107	OPEN	162	12.2		
477	2252.6+3135	OPEN	119	11.3		
478	2255.8+1350	OPEN	230	14.7		
479	2256.5+1933	MED COMP	200	18.2		
480	2256.8+2445	OPEN	169	16.3		
481	2259.6+0746	OPEN	185	14.9		
482	2307.6+0713	COMPACT	175	6.1		
483	2312.5-0229	MED COMP	153	12.8		
484	2315.5+2854	OPEN	165	12.3		
485	2316.5+0046	OPEN	181	10.3		
486	2318.0+1910	OPEN	275	19.2	A2572	
487	2320.0+0845	MED COMP	969	33.8	PEGASUS	
488	2320.2+4309	OPEN	144	10.0		
489	2322.4+1427	MED COMP	352	10.4	A2593	
490	2324.7+3145	OPEN	189	14.4		
491	2329.0-0224	OPEN	235	13.8		
492	2332.8+2027	MED COMP	216	7.1	A2625	A2626
493	2335.5+2449	MED COMP	970	32.3	A2634	A2666
494	2343.4+0845	MED COMP	245	11.1	A2657	
495	2347.5+0707	OPEN	120	16.0		
496	2349.0+1045	OPEN	142	8.4		
497	2352.1+4718	OPEN	178	17.2		
500	1224.0+1320	MED COMP	1000	64.3	VIRGO	

References

and Key to Reference Codes

```
A   1     ARP,H.C.
              1967      APJ     148,    321.
A   2     ARP,H.C.
              1968      APJ     152,   1101.
A   3     ARP,H.C.
              1968      PASP     80,    129.
A   4     ARP,H.C.
              1969      A&A       3,    418.
A   5     ARP,H.C.
              1970      APLET     5,    257.
A   6     ARP,H.C.
              1970      APLET     7,    221.
A   7     ARP,H.C.
              1971      APLET     9,      1.
A   8     ARP,H.C.;KHACHIKIAN,E.YE.
              1973      AFZ       9,    509.
A   9     ARP,H.C.;KHACHIKIAN,E.YE.
              1974      AFZ      10,    173.
A  10     ARP,H.C.;KHACHIKIAN,E.YE.
              1974      AFZ      10,    625.
A  11     ARP,H.C.;O'CONNELL,R.W.
              1975      APJ     197,    291.
AD  1     ARAKELIAN,M.A.;DIBAY,E.A.;YESIPOV,V.F.;MARKARIAN,B.E.
              1970      AFZ       6,     39.
AD  2     ARAKELIAN,M.A.;DIBAY,E.A.;YESIPOV,V.F.;MARKARIAN,B.E.
              1970      AFZ       6,    357.
AD  3     ARAKELIAN,M.A.;DIBAY,E.A.;YESIPOV,V.F.;MARKARIAN,B.E.
              1971      AFZ       7,    177.
AD  4     ARAKELIAN,M.A.;DIBAY,E.A.;YESIPOV,V.F.
              1972      AFZ       8,     33.
AD  5     ARAKELIAN,M.A.;DIBAY,E.A.;YESIPOV,V.F.
              1972      AFZ       8,    177.
AD  5     ARAKELIAN,M.A.;DIBAY,E.A.;YESIPOV,V.F.
              1972      AFZ       8,    329.
AD  6     ARAKELIAN,M.A.;DIBAY,E.A.;YESIPOV,V.F.
              1973      AFZ       9,    319.
AD  7     ARAKELIAN,M.A.;DIBAY,E.A.;YESIPOV,V.F.
              1973      AFZ       9,    325.
AD  8     ARAKELIAN,M.A.;DIBAY,E.A.;YESIPOV,V.F.
              1975      AFZ      11,     15.
AD  9     ARAKELIAN,M.A.;DIBAY,E.A.;YESIPOV,V.F.
              1975      AFZ      11,    377.
AD 10     ARAKELIAN,M.A.;DIBAY,E.A.;YESIPOV,V.F.
              1976      AFZ      12,    195.
AE        ABELL,G.O.;EASTMOND,T.S.;JENNER,D.C.
              1978      APJ L   221,      1.
```

```
AL 1     ALLEN,R.J.
            1970      A&A      7,    330.
AL 2     ALLEN,R.J.;DARCHY,B.F.;LAUQUE,R.
            1971      A&A     10,    198.
AL 3     ALLEN,R.J.;VANDERHULST,J.M.;GOSS,W.M.;HUCHTMEIER,W.
            1978      A&A     64,    359.
AN 1     ANDERSON,K.S.;KRAFT,R.P.
            1969      APJ    158,    859.
AN 2     ANDERSON,K.S.
            1973      APJ    182,    369.
AN 3     ANDERSON,K.S.
            1974      APJ    189,    195.
AS       ANDRILLAT,Y.;SWINGS,J.P.
            1977      APLET   18,    151.
BA 1     BALKOWSKI,C.;BOTTINELLI,L.;GOUGUENHEIM,L.;HEIDMANN,J.
            1973      A&A     23,    139.
BA 2     BALKOWSKI,C.;BOTTINELLI,L.;GOUGUENHEIM,L.;HEIDMANN,J.
            1972      A&A     21,    303.
BA 3     BALKOWSKI;BOTTINELLI;CHAMARAUX,P.;GOUGUENHEIM;HEIDMANN.
            1973      A&A     25,    319.
BA 4     BALKOWSKI;BOTTINELLI;CHAMARAUX,P.;GOUGUENHEIM;HEIDMANN.
            1974      A&A     34,    439.
BB 1     BURBIDGE,E.M.;BURBIDGE,G.R.
            1959      APJ    129,    271.
BB 2     BURBIDGE,E.M.;BURBIDGE,G.R.
            1959      APJ    130,     12.
BB 3     BURBIDGE,E.M.;BURBIDGE,G.R.;PRENDERGAST,K.H.
            1959      APJ    130,     26.
BB 4     BURBIDGE,E.M.;BURBIDGE,G.R.
            1959      APJ    130,    629.
BB 5     BURBIDGE,E.M.;BURBIDGE,G.R.;PRENDERGAST,K.H.
            1959      APJ    130,    739.
BB 6     BURBIDGE,E.M.;BURBIDGE,G.R.;PRENDERGAST,K.H.
            1960      APJ    131,    282.
BB 7     BURBIDGE,E.M.;BURBIDGE,G.R.;PRENDERGAST,K.H.
            1960      APJ    131,    549.
BB 8     BURBIDGE,E.M.;BURBIDGE,G.R.
            1960      APJ    132,     30.
BB 9     BURBIDGE,E.M.;BURBIDGE,G.R.;PRENDERGAST,K.H.
            1960      APJ    132,    640.
BB10     BURBIDGE,E.M.;BURBIDGE,G.R.;PRENDERGAST,K.H.
            1960      APJ    132,    654.
BB11     BURBIDGE,E.M.;BURBIDGE,G.R.;PRENDERGAST,K.H.
            1960      APJ    132,    661.
BB12     BURBIDGE,E.M.;BURBIDGE,G.R.
            1961      AJ      66,    544.
BB13     BURBIDGE,E.M.;BURBIDGE,G.R.
            1961      APJ    133,    726.
BB14     BURBIDGE,E.M.;BURBIDGE,G.R.;PRENDERGAST,K.H.
            1961      APJ    133,    814.
BB15     BURBIDGE,E.M.;BURBIDGE,G.R.;PRENDERGAST,K.H.
            1961      APJ    134,    232.
BB16     BURBIDGE,E.M.;BURBIDGE,G.R.;PRENDERGAST,K.H.
            1961      APJ    134,    237.
BB17     BURBIDGE,E.M.;BURBIDGE,G.R.
            1961      APJ    134,    244.
BB18     BURBIDGE,E.M.;BURBIDGE,G.R.
            1961      APJ    134,    248.
BB19     BURBIDGE,E.M.;BURBIDGE,G.R.;PRENDERGAST,K.H.
            1961      APJ    134,    874.
BB20     BURBIDGE,E.M.;BURBIDGE,G.R.
            1962      APJ    135,    366.
BB21     BURBIDGE,E.M.;BURBIDGE,G.R.;PRENDERGAST,K.H.
            1962      APJ    136,    119.
BB22     BURBIDGE,E.M.;BURBIDGE,G.R.;PRENDERGAST,K.H.
            1962      APJ    136,    128.
BB23     BURBIDGE,E.M.;BURBIDGE,G.R.;PRENDERGAST,K.H.
            1962      APJ    136,    339.
BB24     BURBIDGE,E.M.;BURBIDGE,G.R.;PRENDERGAST,K.H.
            1962      APJ    136,    704.
BB25     BURBIDGE,E.M.;BURBIDGE,G.R.;PRENDERGAST,K.H.
            1963      APJ    137,    376.
BB26     BURBIDGE,E.M.;BURBIDGE,G.R.;PRENDERGAST,K.H.
            1963      APJ    137,   1022.
BB27     BURBIDGE,E.M.;BURBIDGE,G.R.;PRENDERGAST,K.H.
            1963      APJ    138,    375.
BB28     BURBIDGE,E.M.;BURBIDGE,G.R.;HOYLE,F.
            1963      APJ    138,    873.
BB29     BURBIDGE,E.M.;BURBIDGE,G.R.
            1963      APJ    138,   1306.
BB30     BURBIDGE,E.;BURBIDGE,G.;CRAMPIN,J.;RUBIN,V.;PRENDERGAST,K.
            1964      APJ    139,    539.
BB31     BURBIDGE,E.;BURBIDGE,G.;CRAMPIN,J.;RUBIN,V.;PRENDERGAST,K.
            1964      APJ    139,   1058.
BB32     RUBIN,V.C.;BURBIDGE,E.M.;BURBIDGE,G.R.
            1964      APJ    140,     94.
BB33     BURBIDGE,E.M.;BURBIDGE,G.R.;RUBIN,V.C.
            1964      APJ    140,    942.
BB34     BURBIDGE,E.M.;BURBIDGE,G.R.
            1964      APJ    140,   1445.
```

BB35   BURBIDGE,E.M.;BURBIDGE,G.R.;PRENDERGAST,K.H.
              1964   APJ   140, 1617.
BB36   RUBIN,V.;BURBIDGE,E.;BURBIDGE,G.;CRAMPIN,J.;PRENDERGAST,K.
              1965   APJ   141, 759.
BB37   RUBIN,V.;BURBIDGE,E.;BURBIDGE,G.;PRENDERGAST,K.
              1965   APJ   141, 885.
BB38   BURBIDGE,E.M.;BURBIDGE,G.R.
              1965   APJ   142, 634.
BB39   BURBIDGE,E.M.;BURBIDGE,G.R.;PRENDERGAST,K.H.
              1965   APJ   142, 641.
BB40   BURBIDGE,E.M.;BURBIDGE,G.R.;PRENDERGAST,K.H.
              1965   APJ   142, 649.
BB41   BURBIDGE,E.M.;BURBIDGE,G.R.
              1965   APJ   142, 1351.
BB42   BURBIDGE,E.M.;BURBIDGE,G.R.;SHELTON,J.W.
              1967   APJ   150, 783.
BB43   BURBIDGE,E.M.;HODGE,P.M.
              1971   APJ   166,   1.
BB44   BURBIDGE,E.M.;BURBIDGE,G.R.;SOLOMON,P.M.;STRITTMATTER,P.A.
              1971   APJ   170, 233.
BB45   BURBIDGE,E.M.;BURBIDGE,G.R.
              1972   APJ   172,   37.
BC   BARBON,R.;CAPACCIOLI,M.
              1974   A&A   35, 151.
BC   BARBON,R.;CAPACCIOLI,M.
              1976   A&A   49, 125.
BE 1   BERTOLA,F.
              1965   COAA   172,   .
BE 2   BERTOLA,F.
              1966   COAA   186,   .
BF   BRACCESI,A.;FORMIGGINI,L.;GIOIA,I.;SARGENT,W.L.W.
              1972   PASP   84, 592.
BG 1   BOTTINELLI,L.;GOUGUENHEIM,L.
              1976   A&A   47, 381.
BG 2   BOTTINELLI,L.;CHAMARAUX,P.;GOUGUENHEIM,L.;HEIDMANN,J.
              1973   A&A   29, 221.
BG 3   BOTTINELLI;CHAMARAUX;GOUGUENHEIM;HEIDMANN;LAUQUE,R.
              1970   A&A   6, 453.
BG 4   BOTTINELLI,L.;GOUGUENHEIM,L.;HEIDMANN,J.
              1973   A&A   22, 281.
BG 5   BOTTINELLI,L.;GOUGUENHEIM,L.
              1975   A&A   39, 341.
BG 6   BOTTINELLI,L.;DUFLOT,R.;GOUGUENHEIM,L.;HEIDMANN,J.
              1975   A&A   41,   61.
BG 7   BOTTINELLI,L.;GOUGUENHEIM,L.
              1977   A&A   54, 641.
BG 8   BOTTINELLI,L.;GOUGUENHEIM,L.
              1977   A&A L   60,   23.
BG 9   TULLY;BOTTINELLI;FISHER;GOUGUENHEIM;SANCISI;VANWOERDEN.
              1978   A&A   63,   37.
BG10   BOTTINELLI,L.;DUFLOT,G.;GOUGUENHEIM,L.
              1978   A&A   63, 363.
BH   BORCHKHADZE,T.M.
              1974   AFZ   10, 493.
BI 1   BIEGING,J.H.;BIERMANN,P.
              1977   A&A   60, 361.
BI 2   BIEGING,J.H.
              1978   A&A   64,   23.
BJ   BRANDT,J.C.
              1965   MN   129, 309.
BK   BRUNDAGE,W.D.;KRAUS,J.D.
              1966   SCI   153, 411.
BL   BOSMA,A.;EKERS,R.D.;LEQUEUX,J.
              1977   A&A   57,   97.
BM   BLACKMAN,C.P.
              1977   MN   178,   15.
BN   BURBIDGE,E.M.;SARGENT,W.L.W.
              1971   NG   351.
BO   BERTOLA,F.;D'ODORICO,S.
              1973   APLET   13, 161.
BP 1   BERTOLA,F.;CAPACCIOLI,M.
              1977   APJ   211, 697.
BP 2   BERTOLA,F.;CAPACCIOLI,M.
              1978   APJ   219, 404.
BR 1   BARBON,R.
              1969   CKPNO   436,   .
BR 2   BARBON,R.
              1969   CKPNO   510,   .
BS   BURBIDGE,E.M.;STRITTMATTER,P.A.
              1972   APJ L   172,   37.
BS   BURBIDGE,G.R.;O'DELL,S.L.;STRITTMATTER,P.A.
              1972   APJ   175, 601.
BS   STRITTMATTER,P.A.;CARSWELL,R.F.;GILBERT,G.;BURBIDGE,E.M.
              1974   APJ   190, 509.
BT   BARBIERI,C.;BERTOLA,F.;DITULLIO,G.
              1974   A&A   35, 463.
BU   BURBIDGE,E.M.
              1967   APJ L   149,   51.
BU   BURBIDGE,E.M.
              1970   APJ L   160,   33.

```
BV        BOSMA,A.;VANDERHULST,J.M.;SULLIVAN,W.T.
             1977      A&A       57,   373.
BW        BARBIERI,C.;DISEREGO ALIGHIERI,S.;ZAMBON,M.
             1977      A&A       57,   353.
BX        BOKSENBERG,A.;NETZER,H.
             1977      APJ      212,    37.
BY        DEBRUYN,A.G.
             1977      A&A       54,   491.
BZ        BAUTZ,L.P.
             1972      AJ        77,   331.
CA        CAROZZI,N.;CHAMARAUX,P.;DUFLOT-AUGARDE,R.
             1974      A&A       30,    21.
CA        CAROZZI,N.;CHAMARAUX,P.;DUFLOT-AUGARDE,R.
             1974      A&A       33,   113.
CA        CAROZZI,N.
             1976      A&A       49,   425.
CA        CAROZZI,N.
             1976      A&A       49,   431.
CA        CAROZZI,N.
             1977      A&A       55,   261.
CC        CARRANZA,G.J.;CRILLON,R.;MONNET,G.
             1969      A&A        1,   479.
CE        COOKE,J.;EMERSON,D.;NANDY,K.;REDDISH,V.;SMITH,M.
             1977      MN       178,   687.
CH        CHROMEY,F.R.
             1973      A&A       29,    77.
CK        O'CONNELL,R.W.;KRAFT,R.P.
             1972      APJ      175,   335.
CM        CHINCARINI,G.;MARTINS,D.
             1975      APJ      196,   335.
CN        CARRANZA,G.J.;COURTES,G.;GEORGELIN,Y.;MONNET,G.;POURCELOT,A.
             1968      ADA       31,    63.
CR 1      CHINCARINI,G.;ROOD,H.J.
             1971      APJ      168,   321.
CR 2      CHINCARINI,G.;ROOD,H.J.
             1972      AJ        77,     4.
CR 3      CHINCARINI,G.;ROOD,H.J.
             1972      AJ        77,   448.
CR 4      CHINCARINI,G.;ROOD,H.J.
             1976      APJ      206,    30.
CR 5      CHINCARINI,G.;ROOD,H.J.
             1976      PASP      88,   388.
CR 6      CHINCARINI,G.;ROOD,H.J.
             1977      APJ      214,   351.
CT        CRILLON,R.;MONNET,G.
             1969      A&A        1,   449.
CX        CHAMARAUX,P.
             1977      A&A       60,    67.
D         DAVIES,R.D.
             1973      MN    P  161,    25.
DA 1      DUFLOT-AUGARDE,R.
             1961      CMRN     253,   224.
DA 2      DUFLOT,R.
             1965      JOBS      48,   247.
DB        BEALE,J.S.;DAVIES,R.D.
             1969      NAT      221,   531.
DC        DISNEY,M.J.;CROMWELL,R.H.
             1971      APJ L    164,    35.
DD        DEAN,J.F.;DAVIES,R.D.
             1975      MN       170,   503.
DE        DIBAY,E.A.;YESIPOV,V.F.
             1968      AZH       45,   706.
DG        GOTTESMAN,S.T.;DAVIES,R.D.
             1970      MN       149,   263.
DH 1      DEHARVENG,J.M.;PELLET,A.
             1969      A&A        1,   208.
DH 2      DEHARVENG,J.M.;PELLET,A.
             1970      A&A        7,   210.
DH 3      DEHARVENG,J.M.
             1971      THES       ,      .
DJ        DEJAGER,G.;DAVIES,R.D.
             1971      MN       153,     9.
DL 1      DAVIES,R.D.;LEWIS,B.M.
             1973      MN       165,   231.
DL 2      LEWIS,B.M.;DAVIES,R.D.
             1973      MN       165,   213.
DM        DICKENS,R.J.;MOSS,C.
             1976      MN       174,    47.
DM        MOSS,C.;DICKENS,R.J.
             1977      MN       178,   701.
DN        DANZIGER,I.J.;CHROMEY,F.R.
             1972      APLET     10,    99.
DO        D'ODORICO,S.
             1970      APJ      160,     3.
DP 1      DUPUY,D.L.
             1968      PASP      80,    29.
DP 2      DUPUY,D.L.
             1969      PASP      81,   637.
DP 3      DUPUY,D.L.
             1970      AJ        75,  1143.
```

```
DR       GOTTESMAN,S.T.;DAVIES,R.D.;REDDISH,V.C.
         1966       MN       133,  359.
DS 1     DENISYUK,E.K.
         1971       ATS       ,   615.
DS 1     DENISYUK,E.K.
         1971       ATS       ,   621.
DS 2     DENISYUK,E.K.
         1971       ATS       ,   624.
DS 3     DENISYUK,E.K.;LIPOVETSKY,V.A.
         1974       AFZ      10,   315.
DS 4     KOPILOV,I.M.;LIPOVETSKY,V.A.;PRONIK,V.I.;CHUVAEV,K.K.
         1974       AFZ      10,   483.
DS 4     KOPILOV,I.M.;LIPOVETSKY,V.A.;PRONIK,V.I.;CHUVAEV,K.K.
         1976       AFZ      12,   189.
DS 5     DENISJUK,E.K.
         1974       ATS       ,   797.
DS 6     DENISJUK,E.K.
         1974       ATS       ,   809.
DS 7     DENISJUK,E.K.
         1974       ATS       ,   837.
DU       MCCUTCHEON,W.H.;DAVIES,R.D.
         1970       MN      150,   337.
EA       EASTMOND,T.S.;ABELL,G.O.
         1978       PASP     90,   367.
F        FAIRALL,A.P.
         1971       MN      153,   383.
FD       FABER,S.M.;DRESSLER,A.
         1977       AJ       82,   187.
FR 1     RUBIN,V.C.;FORD,W.K.
         1967       PASP     79,   322.
FR 2     FORD,W.K.;PURGATHOFER,A.T.;RUBIN,V.C.
         1968       APJ L   153,    39.
FR 3     RUBIN,V.C.;FORD,W.K.
         1968       APJ     154,   431.
FR 4     FORD,W.K.;RUBIN,V.C.
         1968       PASP     80,   466.
FR 5     BERTOLA,F.;D'ODORICO,S.;FORD,W.K.;RUBIN,V.C.
         1969       APJ L   157,    27.
FR 6     RUBIN,V.C.YFORD,W.K.
         1970       APJ     159,   379.
FR 7     RUBIN,V.C.YFORD,W.K.
         1970       APJ     170,    25.
FR 8     FORD,W.K.;RUBIN,V.C.;ROBERTS,M.S.
         1971       AJ       76,    22.
FR 9     RUBIN,V.C.;THONNARD,N.;FORD,W.K.
         1975       APJ     199,    31.
FR10     RUBIN,V.C.;FORD,W.K.;PETERSON,C.J.
         1975       APJ     199,    39.
FR11     RUBIN,V.C.;FORD,W.K.;THONNARD,N.;ROBERTS,M.S.
         1976       AJ       81,   687.
FR12     KINMAN,T.D.;RUBIN,V.C.;THONNARD,N.;FORD,W.K.;PETERSON,C.J.
         1977       AJ       82,   879.
FR13     PETERSON,C.J.;RUBIN,V.C.;FORD,W.K.;THONNARD,N.
         1978       APJ     219,    31.
FT       FISHER,J.R.;TULLY,R.B.
         1975       A&A      44,   151.
FT       FISHER,J.R.;TULLY,R.B.
         1977       A&A      54,   661.
FV       FREEMAN,K.C.;DEVAUCOULEURS,G.
         1974       APJ     194,   569.
G        GREENSTEIN,J.L.
         1962       APJ     135,   679.
GD       GUDEHUS,D.H.
         1976       APJ     208,   267.
GO       GOAD,J.W.
         1974       APJ     192,   311.
GR       GORDON,K.M.;REMAGE,N.H.;ROBERTS,M.S.
         1968       APJ     154,   845.
GS       GOTTESMAN,S.T.;WRIGHT,M.C.H.
         1973       APJ     184,    71.
GT       GREGORY,S.A.;THOMPSON,L.A.
         1977       APJ     213,   345.
GU       GOUGUENHEIM,L.
         1969       A&A       3,   281.
GW 1     GUELIN,M.;WELIACHEW,L.
         1969       A&A       1,    10.
GW 2     GUELIN,M.;WELIACHEW,L.
         1970       A&A       7,   141.
GW 3     GUELIN,M.;WELIACHEW,L.
         1970       A&A       9,   155.
GW 4     GUELIN,M.;WELIACHEW,L.
         1978       A&A      65,    37.
GY       GREGORY,S.A.;CONNOLLY,L.P.
         1973       APJ     182,   351.
GY       GREGORY,S.A.
         1975       APJ     199,     1.
GY       GREGORY,S.A.
         1975       PASP     87,   833.
GZ       GREENSTEIN,J.L.;ZWICKY,F.
         1962       PASP     74,    35.
```

HG   HUMASON,M.L.;GOMES,A.M.;KEARNS,C.E.
            1961   PASP   73,   175.
HJ   HUCHRA,J.;THUAN,T.X.
            1977   APJ   216,   694.
HK   HECKATHORN,H.M.
            1972   APJ   173,   501.
HO   HODGE,P.W.
            1974   PASP   86,   645.
HS   HUCHRA,J.;SARGENT,W.L.W.
            1973   APJ   186,   433.
HT   HUCHTMEIER,W.;TAMMANN,G.A.;WENDKER,H.J.
            1977   A&A   57,   313.
HU 1   HUCHTMEIER,W.
            1973   A&A   22,   91.
HU 2   HUCHTMEIER,W.
            1973   A&A   23,   93.
J   JENNER,D.C.
            1974   APJ   191,   55.
K   KINTNER,E.C.
            1971   AJ   76,   409.
KD   KODAIRA,K.
            1971   PASJ   23,   589.
KG 1   KNAPP,G.R.;GALLAGHER,J.S.;FABER,S.M.;BALICK,B.
            1977   AJ   82,   106.
KG 2   KNAPP,G.R.;FABER,S.M.;GALLAGHER,J.S.
            1978   AJ   83,   11.
KH 1   KHACHIKIAN,E.YE.
            1972   AFZ   8,   529.
KH 2   KHACHIKIAN,E.YE.
            1973   AFZ   9,   157.
KI   KIRSHNER,R.P.
            1977   APJ   212,   319.
KK   KAZARIAN,M.A.;KHACHIKIAN,E.YE.
            1974   AFZ   10,   477.
KR   KARACHENTSEV,I.D.;PRONIK,V.I.;CHUVAEV,K.K.
            1975   A&A   41,   375.
KR   KARACHENTSEV,I.D.;PRONIK,V.I.;CHUVAEV,K.K.
            1976   A&A   51,   185.
KS   KRUMM,N.;SALPETER,E.E.
            1977   A&A   56,   465.
KT 1   VANDERKRUIT,P.C.
            1973   APJ   186,   807.
KT 2   VANDERKRUIT,P.C.
            1974   APJ   188,   3.
KT 3   VANDERKRUIT,P.C.
            1974   APJ   192,   1.
KT 4   VANDERKRUIT,P.C.
            1975   APJ   195,   611.
KT 5   VANDERKRUIT,P.C.
            1977   A&A   61,   171.
KZ   KOWAL,C.T.;ZWICKY,F.;SARGENT,W.L.W.;SEARLE,L.
            1974   PASP   86,   516.
L   HUMASON,M.L.;MAYALL,N.U.;SANDAGE,A.R.
            1956   AJ   61,   97.
LA   LAUQUE,R.
            1973   A&A   23,   253.
LB   MAYALL,N.U.;DEVAUCOULEURS,G.
            1962   AJ   67,   363.
LN   ALLOIN,D.
            1973   A&A   27,   433.
LS   LEWIS,B.M.
            1974   MEM   78,   75.
MC   MORTON,D.C.;CHEVALIER,R.A.
            1972   APJ   174,   489.
MC   MORTON,D.C.;CHEVALIER,R.A.
            1972   APJ   179,   55.
MC   RICHSTONE,D.O.;MORTON,D.C.
            1975   APJ   201,   289.
ME   MENG,S.Y.;KRAUS,J.O.
            1966   AJ   71,   170.
MK   MINKOWSKI,R.
            1959   APJ   130,   1028.
MK   MINKOWSKI,R.
            1961   AJ   66,   558.
MK   MINKOWSKI,R.
            1961   4BMSP   3,   249.
MM   MALTBY,P.;MATTHEWS,T.A.;MOFFETT,A.T.
            1963   APJ   137,   153.
MS   MELNICK,J.;SARGENT,W.L.W.
            1977   APJ   215,   401.
MU   MUNCH,G.
            1959   PASP   71,   101.
MW   HUMASON,M.L.;MAYALL,N.U.;SANDAGE,A.R.
            1956   AJ   61,   97.
OK   OKE,J.B.;SARGENT,W.L.W.
            1968   APJ   151,   807.
OK   SHIELDS,G.A.;OKE,J.B.;SARGENT,W.L.W.
            1972   APJ   176,   75.
OP   OSTERBROCK,D.E.;PHILLIPS,M.M.
            1977   PASP   89,   251.

187

```
P          PAGE,T.L.
              1970     APJ    159,   791.
PR         PETERSON,C.J.
              1978     PASP    90,    10.
PS         PETERSON,S.D.;SHOSTAK,G.S.
              1974     AJ      79,   767.
PT         PETERSON,B.A.
              1970     AJ      75,   695.
R  1       ROBERTS,M.S.
              1965     APJ    142,   148.
R  2       ROBERTS,M.S.
              1968     APJ    151,   117.
R  3       ROBERTS,M.S.
              1968     AJ      73,   945.
RB         BURNS,W.R.;ROBERTS,M.S.
              1971     APJ    166,   265.
RC         LOVASICH,J.L.;MAYALL,N.U.;NEYMAN,J.;SCOTT,E.L.
              1961     4BMSP    3,   187.
RG         RODGERS,A.;FREEMAN,K.C.
              1970     APJ L  161,   109.
RO         ROTS,A.H.
              1977     AJ      83,   219.
RR         ROGSTAD,D.H.;ROUGOOR,G.W.;WHITEOAK,J.B.
              1967     APJ    150,     9.
RS         ROGSTAD,D.H.;SHOSTAK,G.S.
              1972     APJ    176,   315.
RT         RUDNICKI,K.;TARRARO,I.
              1969     ACTA    19,   171.
RW         WHITEHURST,R.M.;ROBERTS,M.S.
              1972     APJ    175,   347.
S  1       SHOSTAK,G.S.
              1974     A&A     31,    97.
S  2       SHOSTAK,G.S.
              1974     APJ    187,    19.
S  3       SHOSTAK,G.S.
              1975     APJ    198,   527.
SA 1       SARGENT,W.L.W.
              1970     APJ    159,   765.
SA 2       SARGENT,W.L.W.
              1970     APJ    160,   405.
SA 3       SARGENT,W.L.W.
              1972     APJ    173,     7.
SA 4       SARGENT,W.L.W.
              1973     APJ L  182,    13.
SA 5       KORMENDY,J.;SARGENT,W.L.W.
              1974     APJ    193,    19.
SA 6       SARGENT,W.L.W.
              1968     APJ L  153,   135.
SB         STROMBERG,G.
              1925     APJ     61,   353.
SI         SIMKIN,S.M.
              1972     NAT    239,    43.
SK 1       STOCKTON,A.
              1972     APJ    173,   247.
SK 2       STOCKTON,A.
              1974     APJ    187,   219.
SL         SLINGO,A.
              1974     MN     168,   307.
SN 1       SANDAGE,A.
              1966     APJ    145,     1.
SN 2       SANDAGE,A.
              1967     APJ L  150,   145.
SN 3       SANDAGE,A.
              1972     APJ    173,   485.
SN 4       SANDAGE,A.
              1973     APJ    183,   711.
SN 5       SANDAGE,A.
              1974     APJ    194,   223.
SN 5       SANDAGE,A.;TAMMANN,G.A.
              1975     APJ    197,   265.
SN 6       SANDAGE,A.;TAMMANN,G.A.
              1975     APJ    196,   313.
SN 7       SANDAGE,A.;VISVANATHAN,N.
              1978     APJ    223,   707.
SN 8       SANDAGE,A.
              1978     AJ      83,   904.
SO 1       SAKKA,K.;OKA,S.;WAKUMATSU,K.
              1973     PASJ    25,   153.
SO 2       SAKKA,K.;OKA,S.;WAKUMATSU,K.
              1973     PASJ   250,  3173.
SP         STAUFFER,J.;SPINRAD,H.
              1978     PASP    90,    20.
SR         SCOTT,J.S.;ROBERTSON,J.W.;TARENGHI,M.
              1977     A&A     59,    23.
ST         SCHMIDT,M.
              1965     APJ    141,     1.
SW         SEIELSTAD,G.A.;WHITEOAK,J.B.
              1965     APJ    142,   616.
T          TRITTON,K.P.
              1972     MN     158,   277.
```

```
TG  1      TIFFT,W.G.;GREGORY,S.A.
               1971      PASP      83,    810.
TG  2      TIFFT,W.G.;GREGORY,S.A.
               1973      APJ      181,     15.
TG  2      TIFFT,W.G.;GREGORY,S.A.
               1976      APJ      205,    696.
TI  1      TIFFT,W.G.
               1972      APJ      175,    613.
TI  2      TIFFT,W.G.
               1973      APJ      179,     29.
TI  3      TIFFT,W.G.;JEWSBURY,C.P.;SARGENT,T.A.
               1973      APJ      185,    115.
TI  4      TIFFT,W.G.;HILSMAN,K.A.;CORRADO,L.C.
               1975      APJ      199,     16.
TL         TULLY,R.B.
               1974      APJ S     27,    415.
TO         TOHLINE,J.E.;OSTERBROCK,D.E.
               1976      APJ L    210,    117.
TS         THEYS,J.C.;SPIEGEL,E.A.;TOOMRE,J.
               1972      PASP      84,    851.
TT         TIFFT,W.G.;TARENGHI,M.
               1975      APJ      199,     10.
TU         TURNER,E.L.
               1976      APJ      208,     20.
UL  1      DEMOULIN,M-H.
               1965      CMRN     260,   3287.
UL  2      DEMOULIN,M-H.;BURBIDGE,E.M.;BURBIDGE,G.R.
               1968      APJ      153,     31.
UL  3      DEMOULIN,M-H.
               1969      APJ      156,    325.
UL  4      DEMOULIN,M-H.;TUNG CHAN,U.W.
               1969      APJ      156,    501.
UL  5      DEMOULIN,M-H.
               1969      APJ      157,     69.
UL  6      DEMOULIN,M-H.
               1969      APJ      157,     75.
UL  7      DEMOULIN,M-H.
               1969      APJ      157,     81.
UL  8      BURBIDGE,E.M.;DEMOULIN,M-H.
               1969      APJ L    157,    155.
UL  9      DEMOULIN,M-H.
               1970      APJ L    160,     79.
UL10       ULRICH,M-H.
               1971      APJ      163,    441.
UL11       ULRICH,M-H.
               1972      APJ L    171,     35.
UL12       ULRICH,M-H.
               1972      APJ L    171,     37.
UL13       ULRICH,M-H.
               1973      APJ      181,     51.
UL14       ULRICH,M-H.;KINMAN,T.D.;LYNDS,C.R.;RIEKE,G.H.;EKERS,R.D.
               1975      APJ      198,    261.
UL15       ULRICH,M-H.
               1975      A&A       40,    337.
UL16       ULRICH,M-H.
               1976      APJ      206,    364.
UL17       ULRICH,M-H.;IN COLLA ET AL.
               1975      A&A S     20,      1.
UL18       ULRICH,M-H.
               1978      APJ      221,    422.
V   1      DEVAUCOULEURS,G.;DEVAUCOULEURS,A.
               1960      APJ      131,    265.
V   2      DEVAUCOULEURS,G.;DEVAUCOULEURS,A.
               1961      APJ      133,    405.
V   3      DEVAUCOULEURS,G.;DEVAUCOULEURS,A.
               1961      MEM       68,     69.
V   4      DEVAUCOULEURS,G.;DEVAUCOULEURS,A.
               1963      APJ      137,    363.
V   5      DEVAUCOULEURS,G.;DEVAUCOULEURS,A.
               1967      AJ        72,    730.
V   6      DEVAUCOULEURS,G.
               1967      IAUS      30,
V   7      DEVAUCOULEURS,A.;SHOBBROOK,R.R.;STROBEL,A.
               1976      AJ        81,    219.
V   8      DEVAUCOULEURS,G.;DEVAUCOULEURS,A.
               1976      AJ        81,    595.
W          CHINCARINI,G.;WALKER,M.F.
               1967      APJ      147,    407.
W          WALKER,M.F.;CHINCARINI,G.
               1967      APJ      147,    416.
W          WALKER,M.F.;HAYES,S.
               1967      APJ      149,    481.
W          CHINCARINI,G.;WALKER,M.F.
               1967      APJ      149,    487.
W          WALKER,M.F.
               1968      APJ      151,     71.
WD  1      WEEDMAN,D.W.
               1970      APJ      159,    405.
WD  2      WEEDMAN,D.W.
               1972      APJ      171,      5.
```

```
WE  1     WELIACHEW,L.
                    1969      A&A        3,    402.
WE  2     WELIACHEW,L.
                    1971      PASP      83,    609.
WG        WELIACHEW,L.;GOTTESMAN,S.T.
                    1973      A&A       24,     59.
WH        WHITEOAK,J.B.
                    1972      AUJP      25,    233.
WK  1     WEEDMAN,D.W.;KHACHIKIAN,E.YE.
                    1968      AFZ        4,    587.
WK  2     WEEDMAN,D.W.;KHACHIKIAN,E.YE.
                    1969      AFZ        5,    113.
WK  3     WEEDMAN,D.W.;KHACHIKIAN,E.YE.
                    1971      AFZ        7,    390.
WL        WILLS,D.
                    1967      APJ L    148,     57.
WL        WILLS,D.;WILLS,B.
                    1974      APJ      190,    271.
WL        WILLS,D.;WILLS,B.
                    1976      APJ S     31,    143.
WS        WRIGHT,M.C.H.;SEIELSTAD,G.A.
                    1973      APLET     13,      1.
WW        WARNER,P.J.;WRIGHT,M.C.H.;BALDWIN,J.E.
                    1973      MN       163,    163.
Z   1     ZWICKY,F.
                    1964      APJ      139,    514.
Z   2     ZWICKY,F.
                    1965      CGCG       5,     52.
Z   3     ZWICKY,F.
                    1971      CSCEG      ,       .
Z   4     ZWICKY,F.;IN COSMOVICI,C.B.
                    1974      SSR        ,      15.
ZG        GATES,H.S.;ZWICKY,F.
                    1972      AJ        72,    912.
ZH        ZWICKY,F.;HUMASON,M.L.
                    1960      APJ      132,    627.
ZH        ZWICKY,F.;HUMASON,M.L.
                    1960      APJ      133,    794.
ZH        ZWICKY,F.;HUMASON,M.L.
                    1964      APJ      139,    269.
ZS        ZWICKY,F.;SARGENT,W.L.W.;KOWAL,C.T.
                    1969      PASP      81,    224.
```

DATE DUE
